THE ART OF PRESERVING ALL KINDS OF ANIMAL AND VEGETABLE SUBSTANCES FOR SEVERAL YEARS

LECTOR HOUSE PUBLIC DOMAIN WORKS

HAPPY READING!

THE ART OF PRESERVING ALL KINDS OF ANIMAL AND VEGETABLE SUBSTANCES FOR SEVERAL YEARS

NICOLAS APPERT

ISBN: 978-93-5420-448-7

Published: 1812

LECTOR HOUSE LLP
E-MAIL: lectorpublishing@gmail.com

THE ART OF PRESERVING ALL KINDS OF ANIMAL AND VEGETABLE SUBSTANCES FOR SEVERAL YEARS.

A WORK PUBLISHED BY ORDER OF THE FRENCH MINISTER OF THE INTERIOR, ON THE REPORT OF THE BOARD OF ARTS AND MANUFACTURES,

BY

M. APPERT.

TRANSLATED FROM THE FRENCH.

SECOND EDITION.

1812.

EXPLANATION OF THE PLATE.

Fig. 1. A reel with two iron bars, made use of to double the wire, and cut the doubled wire twice the length required for fixing the corks in the bottles.

Fig. 2. A small machine for twisting the wire one-third of its length after having been doubled by fig. 1.

Fig. 3. An instrument for compressing (and, as it were, biting) the corks three quarters of their length, beginning at the smallest end.

Fig. 4. A stool stuffed with straw, furnished with a wooden stand on which the bottles may be placed to be tied. The same stool will serve to sit on during the corking.

Fig. 5. A hollow block of wood, called a bottle-boot (Casse-Bouteille), within which the bottle is set when it is to be corked. This bottle-boot is furnished with a strong bat for beating in the corks.

Fig. 6. A front and side view of pointed pincers, used for twisting the wire employed to keep on the corks, and for cutting off the superfluous ends of the wire. I make use of flat pincers and scissars for this operation.

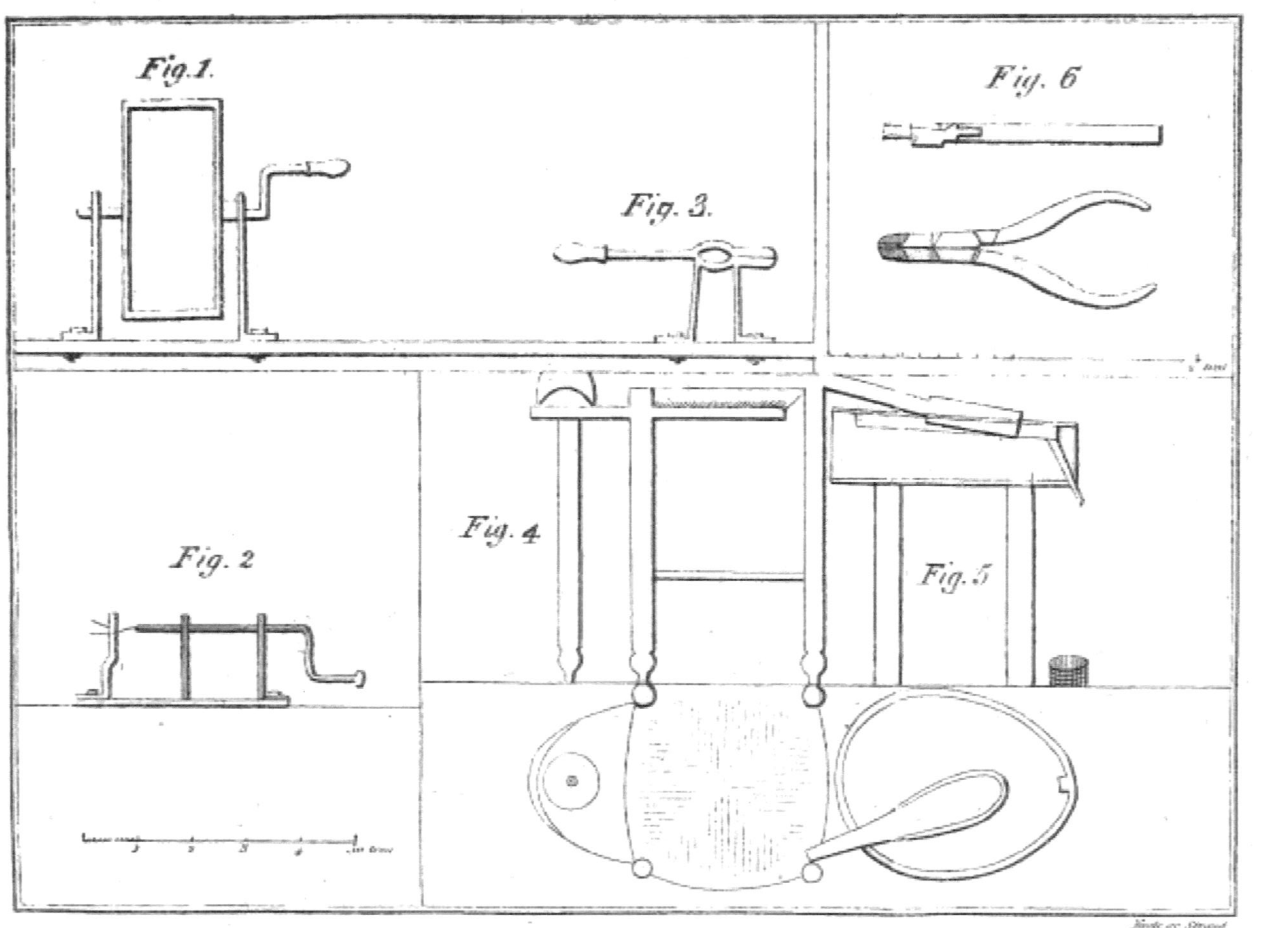

Fig. 1.
Fig. 2.
Fig. 3.
Fig. 4.
Fig. 5.
Fig. 6.

ADVERTISEMENT.

In an advertisement prefixed to the pamphlet, of which the following sheets are a translation, the author publishes his address: "*Quai Napoléon, au coin de la rue de la Colombe, No. 4, dans la Cité, à Paris;*" and offers for sale there, an assortment of provisions, preserved by the process, of which an account is here communicated to the public. As the book itself is a recommendation of the author's own goods, it has been thought proper to add to his account of his process, a translation of the authorities and testimonies by which his own statements are authenticated; notwithstanding the repetitions which are in consequence admitted. The recommendation of the process by the French Minister immediately follows. The more elaborate Report of the Paris Society for the Encouragement of National Industry, will be found at the end of the work.

It is needless to anticipate the author's display of the advantages which must flow from a simple and unexpensive process of keeping fresh articles of animal and vegetable food. If this can be effected for only *one year*, that is, from the season of produce through the seasons of scarcity; if no other articles, for instance, than eggs, cream, and vegetables, can be preserved in their full flavour and excellence during a long winter, there is not a mistress of a family in the kingdom, rich enough to lay by a stock of those articles, and not too rich to despise the economy of a family, who will not find herself benefitted by the perusal of the small work here put within her reach; and there is no reason to suspect the correctness of this part of the author's statements. This, however, is but one of the more obvious benefits of his process; and if thus much be ascertained, then an interminable prospect of resources is opened, which the State, still more than the individual, will be called upon to employ.

The author, in his enumeration of the advantages to be derived from his process, places at the head, the saving it will occasion in the consumption of sugar. This process, added to recent improvement in the art of preparing grape syrup, holds forth, in his opinion, a prospect of relief to the suffering proprietors of French vineyards. This statement will have been listened to with great complacency by the French government, which so ostentatiously avows the determination to compel the whole Continent to subsist on its own produce, and dispense with the more luxurious of transatlantic commodities. Our country, however, from its soil and climate, can take little or no share in this branch of the application of the author's process.

On the other hand it offers us incalculable benefits in the equipment and victualling of our fleets, and in providing for the health and comfort of the floating defence of the country, as well as of that numerous and meritorious class of men, to which the nation owes so much of its prosperity. Whatever promises an improvement in the condition of every order of men who subsist on the Ocean, must be considered as an object of national concern. The French government, at least on the part of some of its members in the subordinate branches of its administration, has taken the lead in recommending the author's process to the attention of public functionaries. From the superior activity, as well as more enlightened discernment of the people of this country, we may expect that our author's process will excite equally the notice of the government and country at large; and we trust that government-contractors and commissioners, as well as the pursers of men of war, and the stewards of merchantmen, will not be the last to examine for themselves the promising statements of our author.

From the public papers we learn that a patent has been taken out for preserving provisions according to the process described in this book. We do not pretend to determine how far this patent may interfere with the adoption by other persons of this same process as a manufactory and trade; but, it is certain, that on the small scale on which provisions would be preserved for single families, every person will be at liberty to avail himself of the instructions he may meet with in this volume.

It was thought less objectionable to insert unnecessary matter, than to omit what to some readers might be useful or interesting. Every thing, therefore, has been translated, and we have even copied the author's plate of the machinery used in corking bottles, though from our improved state of mechanics, the greater part of our readers will stand in no need of its assistance, similar machines being in common use by the wine-coopers, &c.

CONTENTS

LETTER OF THE MINISTER OF THE INTERIOR, COUNT OF THE EMPIRE, TO M. APPERT, &C.

Paris, 30th January 1810.

Second Division.

BOARD OF ARTS AND MANUFACTURES.

My Board of Arts and Manufactures[1] has reported to me, Sir, the examination it has made of your process for the preservation of fruits, vegetables, meat, soup, milk, &c. and from that report no doubt can be entertained of the success of such process. As the preservation of animal and vegetable substances may be of the utmost utility in Sea-voyages, in hospitals and domestic economy, I deem your discovery worthy an especial mark of the good will of the government. I have in consequence acceded to the recommendation made me by my council to grant you a recompence of 12,000 francs.[2] In so doing I had in view the assigning you the reward due to the inventors of useful processes, and also the indemnifying you for the expences you have been obliged to incur, either in the forming your establishment or in the experiments necessary to establish the success of your process. You shall be immediately informed when you may repair to the public treasury and receive the 12,000 francs.

It appears to me of importance, Sir, that you should spread the knowledge of your preserving process. I desire, therefore, that agreeably to your own proposal, you will digest a detailed and exact description of your process. This description, which you will remit to my Board of Arts and Manufactures, shall be printed at your expence, after it shall have been examined. You will then transmit me 200 copies. The transmission of these copies being the only condition I impose on you for the payment of the 12,000 francs, I doubt not you will hasten to fulfil it. I desire, Sir, you will acknowledge the receipt of my letter.

Accept assurances, &c.

(Signed) MONTALIVET.

---

[1] *Mon Bureau consultatif des Arts et Manufactures.*
[2] About £500 sterling.

CERTIFICATE OF THE BOARD OF ARTS

BOARD OF ARTS AND MANUFACTURES.

The undersigned Members of the Board of Arts and Manufactures attached to the Minister of the Interior, being required by his Excellency to examine the description of the process of Mr. Appert for the preservation of alimentary substance, certify that the details it contains, as well on the mode of carrying on the process as on the results, are exactly conformable to the various experiments which Mr. Appert has made before them, by order of his Excellency.

(Signed) BORDEL,
GAY-LUSSAC,
SCIPION-PERRIER,
MOLARD.

Paris, 19th April 1810.

LETTER TO GENERAL CAFFARELLI FROM THE COUNCIL OF HEALTH

Copy of a Letter written to General Caffarelli, Maritime Prefect at Brest, by the Council of Health, dated Brumaire, year 12.

The provisions prepared according to the process of Citizen Appert and sent to this port by the Minister of Marine, have, after lying in the roads three months, been found in the following condition.

The broth or soup (*bouillon*) in bottles was good; the *bouillon* with a *bouilli* in a vessel apart was also good, but weak; the *bouilli* itself was very eatable.

The beans and green peas, prepared both with meat and vegetable soup, had all the freshness and flavour of recently gathered vegetables.

(Signed) Dubreuil,
Billard,
Duret,
Pichon,
Thaumer.

True Copy.

J. Miriel, *Secretary*.

§ I.

THE ART OF PRESERVING, &C. &C. &C.

ALL the expedients hitherto made use of for preserving alimentary and medicinal substances, may be reduced to two principal methods; that of dessication; and that of mingling, in greater or less quantities, a foreign substance for the purpose of impeding fermentation or putrefaction.

It is by the former of these methods that we are furnished with smoaked and hung meat, dried fish, fruits, and vegetables. By the latter, we obtain fruits and other vegetable substances preserved in sugar, the juices and decoctions of plants reduced to syrups and essences, all kinds of pickles, salted meat and vegetables. But each of these modes has its peculiar inconveniences. Dessication takes away the odour, changes the taste of the juices, and hardens the fibrous or pulpy matter (the *porenchyma*).

Sugar, from the strength of its own flavour, conceals and destroys in part other flavours, even that, the enjoyment of which we wish to preserve, such as the pleasant acidity of many fruits. A second inconvenience is this, that a large quantity of sugar is required in order to preserve a small quantity of some other vegetable matter; and hence the use of it is not only very costly, but even in many cases pernicious. Thus the juices of certain plants cannot be reduced to a syrup or essence, but by means of nearly double the quantity of sugar. It results from this, that those syrups or essences contain much more sugar than any medicinal substance, and that most frequently the sugar counteracts the operation of the medicine, and is hurtful to the patient.

Salt communicates an unpleasant acerbity to substances, hardens the animal fibre, and renders it difficult of digestion. It contracts the animal parenchyma.[3] On the other hand, as it is indispensable to remove, by means of water, the greater part of the salt employed; almost all the principles which are soluble in cold water, are lost when the salt is taken away: there remains nothing but the fibrous matter,

---

[3] "The salt meat with which the crews of vessels are fed, appears to be one of the principal causes of the scurvy. It seems that the same causes which operate to prevent the fermentation of meat, renders it also difficult of digestion. Though a small quantity of salt may be an obstacle in the way of putrefaction, the too abundant and frequent use which is made of it, must cause great obstructions in the smaller vessels of the body, and these obstructions cannot fail to overload the stomach of men who have to digest dry vegetables and biscuits, which sailors advanced in years are not always able to chew completely. Bad digestion and obstruction in the smaller vessels may occasionally give rise to ulcers in the mouth, and spots, which denote the scurvy." —*Santé des Marins, by Duhamel.*

or parenchyma; and even that, as has been said, undergoes a change.

Vinegar can seldom be made use of, but in the preparation of certain articles for seasoning.

I shall not enter into any details concerning what has been said and published on the art of preserving alimentary substances. I shall only observe, that as far as my knowledge extends, no author, either ancient or modern, has ever pointed out, or even led to the suspicion, of the principle which is the basis of the method I propose.

It is known, how much, within a certain period, the public attention, both at Paris and in the departments, has been directed towards the means of diminishing the consumption of sugar, by supplying its place by the use of various extracts, or essences, of indigenous substances. The government, whose philanthropic views are turned towards all useful objects, does not cease to invite all those who pursue the arts and sciences, to investigate the means of drawing the utmost advantage from the productions of our soil, in order to develope, to the utmost, our agriculture and manufactures, and so diminish the consumption of foreign commodities.

In order to attain the same end, the Society for the Promotion of National Industry[4] stimulates, by the offer of flattering rewards, all those whose talents and labours are directed towards discoveries, from which the nation and humanity may draw substantial benefits. Animated by this laudable zeal, the Agricultural Society, by its resolution of the 21st of June 1809, and its official notification of it, the 15th of the July following, made an appeal to the whole nation, in order to collect all the information and documents which might contribute to the composition of a work on the art of preserving, by the best possible means, every kind of alimentary substance.

It was after invitations of so great weight, that I resolved to make known a method of effecting this object, of great facility in the execution, and at the same time very cheap, and which, by the extension it admits of, may afford numerous advantages to society.

This method is not a vain theory. It is the fruit of reflection, investigation, long attention, and numerous experiments, the results of which, for more than ten years, have been so surprising, that notwithstanding the proof acquired by repeated practice, that provisions may be preserved two, three, and six years, there are many persons who still refuse to credit the fact.

Brought up to the business of preserving alimentary substance by the received methods; having spent my days in the pantries, the breweries, store-houses, and cellars of Champagne, as well as in the shops, manufactories, and warehouses of confectioners, distillers, and grocers; accustomed to superintend establishments of this kind for forty-five years, I have been able to avail myself, in my process, of a number of advantages, which the greater number of those persons have not possessed, who have devoted themselves to the art of preserving provisions.

I owe to my extensive practice, and more especially to my long perseverance,

---

[4] *La Société d'Encouragement pour l'Industrie nationale.*

the conviction:

1st. That fire has the peculiar property, not only of changing the combination of the constituent parts of vegetable and animal productions, but also of retarding, for many years at least, if not of destroying, the natural tendency of those same productions to decomposition.

2d. That the application of fire in a manner variously adapted to various substances, after having with the utmost care and as completely as possible, deprived them of all contact with the air, effects a perfect preservation of those same productions, with all their natural qualities.

Before I state the details of my process, I ought to observe that it consists principally,

1st. In inclosing in bottles the substances to be preserved.

2d. In corking the bottles with the utmost care; for it is chiefly on the corking that the success of the process depends.

3d. In submitting these inclosed substances to the action of boiling water in a water-bath (BALNEUM MARIAE), for a greater or less length of time, according to their nature, and in the manner pointed out with respect to each several kind of substance.

4th. In withdrawing the bottles from the water-bath at the period described.

§ II.

DESCRIPTION OF MY ROOMS SET APART FOR CARRYING ON THE PROCESS ON A LARGE SCALE.[5]

My laboratory consists of four apartments. The first of these is furnished with all kinds of kitchen utensils, stoves, and other apparatus, necessary for dressing the animal substances to be preserved, as well as with a kettle for broth, gravy, &c. containing 180 French pints, raised on brick work. This kettle is provided with a pot to be put within it, pierced with holes like a skimmer, with divisions for holding various kinds of meat and poultry. This pot can be put into and taken out of the kettle with ease. The kettle is provided with a wide cock, to which is fitted, within, a little rose, like that of a watering-pot, covered with a piece of boulting-cloth. In this way I can procure broth or gravy quite clear, and ready to be put into bottles.

The second apartment is appropriated to the preparing of milk, cream, and whey.

The third is used for corking and tying the bottles and vessels, and putting them into bags.

The fourth is furnished with three large copper boilers, placed upon stones raised on brick work. These boilers are all furnished with a stout lid, fitted, to rest upon the vessels within. Each boiler is furnished with a wide cock below, in order to let out the water at a proper time. These large boilers are destined to receive, generally, all the objects intended to be preserved, in order to apply the action of heat to them in a suitable manner; and thus they constitute so many water-baths.[6]

---

[5] It is obvious, that for the use of private families, and for carrying on the process on a small scale, nothing further will be requisite, than such vessels and other conveniences as are found in every house in the country, where provisions are cured for the consumption of the family during winter.

[6] The reason why it is necessary that large boilers should be furnished with wide cocks is, that it would take up too much time to let so large a body of water, always placed over a heated stove, remain till it became cool; and that, on the other hand, it would do great injury to those substances to let them remain too long exposed to the heat. Without inconvenience, therefore, in private families, any cauldron or earthen vessel may be taken for a water-bath, provided the water rises to the rim of the bottle. In case there should be no vessel sufficiently high, the bottles may be laid down in the water-bath, care being taken to pack them well together, lest they should be broken. Many operations have succeeded well with me this way. The corks are somewhat more liable to burst outwards; but if the bottles are well corked, there is nothing to be feared. For instance, it would not be advisable to

The utensils which furnish the third apartment for the preparatory process consist of

1. Rows of bottle-racks round the room.

2. A reel for the iron wire, to be used for binding the necks of the bottles and other vessels. (*Fig. 1.*)

3. Shears and pincers for tying on the corks. (*Fig. 6.*)

4. Machine for twisting the iron-wire after it has been divided and cut to a proper length. (*Fig. 2.*)

5. Two instruments forming a lever, and used for compressing, and as it were biting the corks. (*Fig. 3.*)

6. A bottle-boot or block, standing on three legs, and provided with a strong bat for corking. (*Fig. 5.*)

7. A stool standing on five legs, for tying on the corks. (*Fig. 4.*)

8. A sufficient quantity of linen bags, for covering the bottles and other vessels.

9. Two stools covered with leather and stuffed with hay, in order to shake the bottles upon them, and in that way force a greater number of peas and other small substances into the bottles.

10. A press for the juice of plants, fruits, and herbs; with pans, vessels, sieves, and every thing else that belongs to it.

Besides my laboratory, consisting of these articles, I have fitted up three apartments.

The first, for preparing vegetables: it is furnished with dressers all round.

The second, for storing up and preparing all kinds of fruit.

The third is a cellar, furnished with bottle racks, for rinsing and setting by the bottles and other vessels, as in a store-house.

I have the precaution to keep the bottles and other vessels I may want, ready rinsed at hand. I am also supplied with an assortment of corks, compressed and bit in the instrument already described. When every preparation is thus made, the process is half done.

The principle by which all alimentary substances are preserved and kept fresh, is invariable in its effects. The result in particular experiments, depends upon the fitness of each individual application of the principle to the substance which is to be preserved, according to its peculiar qualities; but in every case, the exclusion of air is a precaution of the utmost importance to the success of the operation: and in order to deprive alimentary substances of contact with the air, a perfect knowl-

lay on their sides, bottles, or other vessels stopped up with stoppers consisting of different pieces of any substance, because the action of the fire upon this kind of stopper is stronger; and however well the vessel might be corked, it would not be advisable to incur the risque.

Small water-baths are the more convenient, because they may be placed any where, and removed at will. They soon become cold. The bottles are taken out when the water is sufficiently cool to allow of the finger being put in, and thus the operation is terminated.

edge of bottles and the vessels to be used, of corks and corking, is requisite.

§ III.

OF BOTTLES AND VESSELS.

I chose glass, as being the matter most impenetrable by air, and have not ventured to make any experiment with a vessel made of any other substance. The ordinary bottles have generally necks too small and ill made; they are also too weak to resist the blows from the bat and the action of the fire: I, therefore, caused bottles to be made for my especial use, with wider necks, and those necks made with a projecting rim, or ring, on the interior surface, placed below, and resembling, in form, the rim which is at the top of the exterior surface of the necks of bottles. My object was, that when the cork had been forced into the neck of the bottle, three-fourths of its length, in the manner already described, it should be compressed in the middle. In this manner the bottle is perfectly corked on the outside as well as within. It thus opposes an obstacle to the swelling, or expansion, which arises from the operation of heat upon the substance enclosed within the bottle. This mode of forming the neck of the bottle is so much the more indispensable, as I have repeatedly known the swelling to be so strong, as to push out corks of three or four lines in length, though confined by two iron wires crossed. The bottles and vessels should be made of a tough substance [*de matière liante*], the former having the weight of twenty-five or twenty-six ounces for each *litre*[7] that the bottle contains. The glass ought to be of equal thickness in every part, or it is liable to break in the water-bath. The form of the Champagne bottle is most convenient; it is the handsomest as well as the strongest, and is of the best shape for packing up.

---

[7] The French *litre*, consists of nearly two wine pints and a half, English measure.

§ IV.

OF CORKS.

Economy in corks is generally very unwise, as in order to save a very trifle in the price of cork, a risk is incurred of losing the valuable commodity it is intended to preserve. As corking is made use of in order to preserve and meliorate certain articles, by depriving them of all contact with the air, too much attention cannot be given to the good quality of the cork, which should be of eighteen or twenty lines in length and of the finest quality. Experience has so fully satisfied me on this point, that I never make use of any but superfine corks: these are, in the end, the cheapest. I further take the precaution of compressing, and, as it were, biting the cork, three-fourths of its length, by means of the instrument already described (*fig.* 3), beginning at the small end. The cork is rendered more supple; the pores of the cork are brought closer; it is somewhat lengthened, and its thickness is so much diminished at the extremity which is put into the mouth of the bottle, that a large cork may be made to enter a very moderate opening. The action of the heat within the vessel is such, that the cork swells within, and the corking is thus rendered perfect.

§ V.

OF CORKING.

After what has been just said, the absolute necessity will be apparent of having good bottles, with a projecting rim of equal thickness all round within the neck. Excellent superfine corks are also indispensable, which have been compressed in the instrument three quarters of their length.

Before I cork, I take care that the bottles containing liquor are filled only up to within three inches of the outer rim, lest they should burst from the bubbling and swelling occasioned by the application of heat to the water. When the bottles contain vegetables, fruit, &c. they may be filled up to within two inches of the rim.

I place the full bottle upon the bottle-boot already mentioned, before which I seat myself. This apparatus is to be supplied with a strong wooden bat, a small pot full of water, and a sharp knife, greased with a little suet or soap, for cutting off the tops of the corks, which ought never to be raised much above the head of the bottles. These arrangements being made, I place the bottle-boot between my legs, and taking a cork of fit size, I dip one half of it into the little pot of water, in order to facilitate its entrance; and having wiped the end, I then put it to the mouth of the bottle, at the same time turning it round. I hold it in this position with my left hand, which I keep steady, that the bottle may stand upright. I take the bat in my right hand, in order to drive in the cork by force of blows.

When I find, at the first or second blow of the bat, that the cork has somewhat entered, I take my hand from the cork in order to hold with it the neck of the bottle, which I fix firmly and upright upon the bottle-boot; and by dint of repeated blows, I continue to drive in my cork three-fourths of its length. The quarter of the cork which remains above the bottle, after having refused to yield any further to the redoubled blows of the bat, assures me, in the first place, that the bottle is completely corked, and this same residue serves also to hold the double crossed iron wire which is necessary to bind fast the cork, that it may be able to resist the action of heat on the water-bath. I must repeat again, that too much attention cannot be given to the corking: no circumstance however minute ought to be neglected, in order to effect the rigorous exclusion of the air from the substance to be preserved; air being a most destructive agent, and the one which is most sedulously to be counteracted in the course of the process.[8]

[8] Many persons believe they have corked well, when they have forced the cork even with the mouth of the bottle; but this is a great mistake. On the contrary, whenever the whole of the cork, instead of withstanding the blows of the bat, is forced into the bottle, it is adviseable to draw it out and substitute another in its place. Thus the believing that a bottle

The bottles being well stopped up, I then fasten the cork down with a couple of iron wires crossed: this is an easy operation, and any one can do it, who has once seen it done.

I then put each bottle in a bag of canvass or coarse linen cloth, made for the purpose, sufficiently large to wrap up the whole of the bottle up to the very cork. These bags are made in the shape of a muff, open alike at both ends: one of these ends is drawn with a string running in a gutter, leaving an opening of about the width of a crown piece; the other end is provided with a couple of small strings, in order to tye the bag round the neck of the bottle.

By means of these bags, I can dispense with the use of hay or straw in packing up the bottles in the water-bath; and, whenever any one of them breaks, the fragments are preserved in the bag. I am spared a great deal of trouble and a number of inconveniences which I had formerly to sustain, in picking up the pieces of the bottle out of the straw or hay I then made use of.

After having spoken of bottles, their form and quality; of stoppers, and the length of the fine cork of which they ought to be composed; of the corking and tying; of bags, their form and utility; I proceed to give an idea of vessels with large necks, that is, glass jars, which I make use of for preserving solid and bulky substances, such as poultry, game, meat, fish, &c.

These jars have necks of two, three, or four inches diameter, and are of a larger or small size; like bottles, they are furnished with a projecting rim, not only in order to strengthen the neck, but also for receiving the iron wire destined to bind the corks. I have not yet been able to procure from the glass-houses a similar projecting rim in the interior of the neck of these jars, as I have in that of the bottles. The completely corking up these vessels, is, from this circumstance, rendered more difficult, and demands especial care.

I met with another obstacle in the cork itself, from its thinness (more especially when the cork was very fine), and also from its ascending pores being against the grain. I was therefore obliged to form stoppers of three or four pieces of cork, from twenty to twenty-four lines in length, placed together the way of the grain, the corked very low is well corked, because no liquor escapes when the bottle is turned with its neck downwards, is an error, which, joined to the use of bad corks, causes a number of losses. He who corks with care and judgment is satisfied that the operation has been performed well by the resistance of the cork to the blows of the bat, and never thinks of turning the neck of the bottle downwards. It is besides sufficient to reflect on the punctures met with in cork, and on all the hidden defects which may subsist in the interior even of the finest cork, by means of which, the air may be introduced; in order to be convinced of the propriety of making use of none but the very best corks, and that, after having well compressed them in the machine for that purpose; and also of corking them so closely that they become very much compressed in the middle.

It is in this way only, that losses can be prevented from frequently taking place, which have often no other cause than bad corking; for, if a bottle does not instantly run when carelessly corked, it proceeds from this circumstance, that the air has not had time to penetrate through the apertures which may be in the interior of the cork: and, in fact, how different is the quality of wine, drawn from the same cask! and how many bottles do we meet with, which have lost more or less of their contents!

pores of the cork being placed horizontally, by means of isinglass prepared in the following manner.

I melted over the fire four drams of well beaten isinglass, in eight ounces of water: when melted, I caused it to run through fine linen; and then put it again over the fire in order to reduce it to one third of its volume. After which I added an ounce of good full-proof brandy. I then left the whole on the fire till it became reduced to about three ounces. I then put the glue thus prepared in a little pot over live coals, and took care to warm my pieces of cork. I then slightly smeared over the pieces of cork with a brush, in order to glue them together. When the pieces composing the stopper were well fixed and glued together, I then fixed a tight thread to the two extremities of the stopper, in order to keep the pieces together, and let them dry, either in the sun or in a gentle heat for about a fortnight. At the end of this time I took a cork-maker's knife and cut my stoppers of a proper shape; and having always fitted them to the mouth of the jar, they have never proved defective.

Having corked my jars, and driven in the stopper by means of the bat, the bottles being always placed upright in the bottle-boot, I made use of a compound luting. This luting (communicated to me by Mr. Bardel) is made of quick lime, which is slaked in the air by being sprinkled with water, till it becomes reduced to a powder. The powder to be kept in this state in corked bottles ready for use. This lime mixed with a cheese made of skimmed milk (*fromage à la pie*), and formed to the thickness of paste, produces a luting which hardens rapidly, and which withstands the heat of boiling water.

I besmear the whole of the outside of the stopper with this luting, and I cover the edge of the jar with hemp and strips of linen placed above and close to the stopper, and hanging down to the rim.

Farther, that the iron wire may have force enough to keep down the stopper, I put a piece of cork seven or eight lines high, and sixteen or eighteen lines in diameter, in the middle of the large stopper which is itself too big to allow the wire to have any effect upon it. By means of this second cork, placed in the middle of the large stopper, I am able to make the wire take a proper hold of the cork and give due strength and solidity to the stopper.

When every thing has thus been foreseen and prepared, and, above all things, well corked, tied, and wrapped up in bags, there remains nothing to be done, but to apply the preserving principle, that is, *heat*, to the substances duly arranged, and this is the most easy part of the operation.

I place all the vessels, bottles or jars, upright in a boiler, which I then fill with cold water up to the necks of the vessels; I then cover the boiler with its lid, which is made to rest upon the vessels. I cover the upper part of the lid with a piece of wet linen, in order that the sides of the lid may exactly fit, and all evaporation from the water-bath be impeded as much as possible.

When the boiler has been thus filled and adjusted, I light the fire beneath. When the water-bath begins to boil, I take care to maintain the same degree of heat for the greater or less quantity of time required by the substances exposed to its influence. When this time has elapsed, I then instantly put out the fire by means of a coal extinguisher (*étouffoir*).

After the fire has been put out a quarter of an hour, I let out the water of the bath by means of the cock; after the water has been withdrawn half an hour, I uncover the boiler, and I do not take out the bottles till one or two hours after the uncovering; and this terminates the operation.

The next day, or a fortnight afterwards, for that is immaterial, I place my bottles on shelves, as I do wine, in a cool and shady place. If I purpose sending them a great distance, I think it worth while to pitch them before I place them on the shelves; otherwise this last operation is not absolutely necessary. I have now by me, bottles which have been three years lying under a stair-case, the substances contained in which retain as much flavour as if they were just prepared, and yet they were never pitched.

We have just seen, from all that has been said, that alimentary substances, in order to be preserved, should be, without exception, subjected to the application of heat in a water-bath; after being rigorously excluded from all contact with the air, in the manner, and with the precautions already indicated.

The preserving principle is, as I have already observed, invariable in its effects. Thus every loss I have sustained from any of the articles being spoiled, had no other cause than an erroneous application of the principle, or some negligence or omission in the preparatory process already pointed out. It sometimes happens to me even now, that my operations do not perfectly succeed; but no man makes experiments in any of the arts, or in any branch of natural philosophy, without being liable to disappointment. Nor can any one, therefore, who is employed in such a process as mine, flatter himself that he may not sometimes find his commodities spoiled from some defect in a vessel, or in the interior of a cork. But in fact, when due attention is given, these losses seldom take place.

§ VI.

THE MEANS OF DISTINGUISHING AMONG THE BOTTLES OR JARS, AS THEY ARE TAKEN FROM THE BOILER, SUCH OF THEM AS, FROM SOME NEGLECT IN THE PREPARATORY PROCESS, SOME ACCIDENT, OR THE ACTION OF THE FIRE, ARE IN DANGER OF OCCASIONING A LOSS, OR SPOILING THE SUBSTANCES ENCLOSED IN THEM.

When the operation is completed, of whatever kind it may be, I take the greatest care in my power to examine all the bottles and jars one by one, as I take them from the boiler.

I have remarked in some, defects in the glass, as stars and cracks occasioned by the action of the heat in the water-bath; or by the tying, when the mouth of the vessel has been too weak.

I have observed in others, a moisture round the stopper, or little spots near the mouth, from which I inferred that part of the substance enclosed had oozed out during the dilation or expansion produced by the heat of the water-bath: these are the two principal observations that usually occur to me: and whenever I observe either of these appearances on any bottle, I always set it aside, and make use of the substance immediately, that nothing may be lost.

The first of the flaws pointed out, arises from the quality, and originally bad structure of the bottle; but the second may arise from any one of four causes:—1. From a bad cork; 2. from bad corking; 3. from the bottle having been filled too near the brim; and 4. from bad tying. A single one of these faults is sufficient to spoil a bottle; more easily, therefore, a complication of them.

In the applying of heat to the water-bath, I have had various obstacles to encounter, more particularly when peas were to be preserved; for peas are of all substances the most difficult to preserve completely. This vegetable, when gathered while it is too young or too tender, dissolves in water, and in consequence the bottle is found half empty, and even this half is not fit to be kept; hence, whenever this circumstance occurs, I set aside the bottle and make use of the article immediately. If the peas have been gathered two or three days, the heat occasions them to lose all their flavour; they become hard; they ferment before the operation; the bottles break in the water-bath with an explosion; those which resist the first heat break afterwards, or are faulty: and this is easily recognised by the liquor in the bottle, which becomes turbid; while peas which are well preserved, leave the liquid pellucid.

It is not necessary to recommend dispatch and the utmost cleanliness in the preparation of alimentary substances. This is absolutely indispensable; more especially in what respects the substances themselves, which are to be preserved.

I take care to have all my preparatory arrangements made before I begin the process; that there may be no waiting, and that the best use may be made of the time employed in carrying it on.

DESCRIPTION OF MY PROCESS, AS APPLIED TO THE VARIOUS ARTICLES INTENDED TO BE PRESERVED.

§ VII.

BOILED MEAT.
(POT-AU-FEU DE MÉNAGE.)

I put a quantity of meat into the pot to be boiled in the ordinary way. When it was three-fourths boiled, I took out one half of it, the bones of which I had already taken off, as I purposed to preserve it. When the meat was completely boiled, I strained the broth, and after it had become cool, I put it in bottles which I corked well, tied and wrapped up in their several bags. The beef which I had taken out when three-fourths done, I put into jars which I filled up with a part of the same broth. Having corked, luted, and tied up these, and wrapped them in bags, I placed them, and the bottles containing the broth, upright in a cauldron or boiler. I filled this boiler with cold water up to the rim of the bottles and jars. I put the lid upon the boiler, causing it to rest on the vessels within, and took care to surround it with a wet linen cloth, in order to impede as much as possible, any evaporation from the water-bath. I heated the boiler, and when the water-bath had been made to boil, I kept up the same degree of heat for an hour, and precisely at the end of the hour, let the fire pass into an extinguisher. Half an hour afterwards, I let off the water from the bath, by means of the cock at the bottom of the boiler. At the end of another half hour I took off the lid. An hour or two afterwards, I took out the bottles and jars. (The time of doing this is, however, immaterial, and the operator will consult his own convenience.) The next day I besmeared the corks with rosin, in order to forward the bottles and jars to different sea-ports.

At the end of a year, and a year and half, the broth and boiled meat were found as good as if made the day they were eaten.

§ VIII.

GRAVY.

In the year 12, having reason to hope that I should be employed to provide some nourishing provisions for the sick on board his majesty's vessels, in consequence of some experiments which had already been made in the sea-ports, by order of *his Excellency the Minister of the Marine and Colonies*, on alimentary productions preserved according to my method; I made the necessary arrangement for fulfilling the orders I had reason to expect. In consequence, that I might not want too many bottles and jars, and that I might be able to condense the substance of eight messes in a bottle of the size of one *litre*, I made the following experiment. As, in general, evaporation cannot take place, but at the expence of the object to be condensed,[9] I made some gravy, in the proportion of two pounds of good meat and poultry to one *litre*. My gravy being made, and strained and suffered to become cool, I put it in bottles. After having well corked, and tied the bottles and wrapped them in bags, I placed them in the boiler. I had taken out, when one quarter dressed, the best pieces of the beef and poultry. When these were grown cold, I put them in jars, and filled the jars with the same gravy. Having well corked, luted, tied and wrapped up these jars, I set them upright in the same boiler with the bottles of gravy. Having filled the boiler with cold water up to the rim of the bottles and jars, and having covered the lid of the boiler with a wet linen cloth, I heated the water-bath. When it was made to boil I kept up the same degree of heat for two hours, and completed this operation as I did the preceding.

The beef and fowls were found well dressed, and were kept, as well as the gravy, for more than two years.

---

[9] Jellies, essences of meat, the substance of ice and portable soups, which are prepared from the soft and white parts of animals, preserved at a great expence by means of evaporation, and drying in stoves with the aid of hartshorn and isinglass, furnish merely factitious aliments, without flavour or any other than a burnt or mouldy taste.

§ IX.

BROTH, OR JELLY.

I composed this jelly, according to the prescription of a physician, of calves feet and lights, red cabbage, carrots, turnips, onions, and leeks, taking a sufficient quantity of each. A quarter of an hour before I took this jelly from the fire, I added some sugar-candy with some Senegal gum. I strained it as soon as it was made. After it was cold it was put in bottles, which were corked, tied, wrapped up in bags, and put in the water-bath, which was kept boiling one quarter of an hour, and this jelly was preserved and remained as good as it was the day on which it was made.

§ X.

ROUND OF BEEF, FILLET OF MUTTON, FOWLS AND YOUNG PARTRIDGES.

I prepared all these articles as if for common use, but only three-fourths dressed, the young partridges being roasted. When they were grown cold, I put these articles separately into jars of a sufficient size. Having well corked, luted, tied and wrapped them up, I put them all into the water-bath which was kept on the boil for half an hour. They were forwarded to Brest, and from thence were sent to Sea for four months and ten days, together with some vegetables, gravy, and preserved milk, all well packed up in a chest.

When opened, eighteen different kinds of preserved food were tasted, every one of which had retained its freshness; and not a single substance had undergone the least change at Sea.

To the experiments made with these four kinds of provisions, I can add two others made by myself; the one, a fricasee of fowls; and the other, a matelot of eels, carp, and pike, with an addition of sweet-bread, mushrooms, onions, butter, and anchovies, all dressed in white wine. The fricasee and the matelot were perfectly preserved.

These results prove sufficiently that the same principle, applied with the same preparatory process, and with the same care and precautions, in general preserves all animal productions. But it is to be observed that in the previous cooking of each articles, it is to be only three-fourths dressed at the utmost, in order that the remainder of the requisite cooking may be communicated by means of the water-bath.

There are a number of articles which can bear an additional hour of boiling in the water-bath without any danger, as broth, gravy, jellies and the essences of meat, poultry and ham, the juice of the grape and of plants, &c. But there are also others which will sustain a great injury from a quarter of an hour's or even a minute's too much boiling. Thus the result will always depend upon the dexterity, intelligence and judgment of the operator.[10]

---

[10] The celebrated CHAPTAL says, in his *Elémens de Chimie, discours préliminaire*, p. cxxxi. "We hear in manufactories of nothing but the *caprice of experiment*, but this vague phrase has its origin only in the ignorance in which the workmen are of the true principles of their art; for nature does not act according to any principle of discernment, but obeys constant laws. The dead matter which we employ in our manufactories, exhibits necessary effects in which the will can have no share, and consequently can have no caprice. Make yourself acquaint-

§ XI.

NEW-LAID EGGS.

The more fresh the egg is, the longer it withstands the heat of the water-bath. I consequently took eggs the day they had been laid, placed them in a jar, with raspings of bread, to fill up the vacuities, and secure them against breaking when removed to a distance. Having well corked, tied and luted the jars, &c. I placed them in a boiler of a proper size[11] to give them seventy-five degrees of heat.[12] Having taken the water-bath from the fire, I took out the eggs as soon as the water was so cool that I could put my finger in it. I then took out the eggs and kept them six months. At the end of that period I took the eggs out of the jar, put them into cold water which I set on the fire, and heated it to seventy-five degrees: I found them fit to dip a toast of bread into, and as fresh as when I prepared them. As to hard eggs, which are to be cut into slices and fricaseed, I heat the water-bath eighty degrees, and as soon as it begins to boil, I remove the water-bath from the fire.

---

ed, we should say to the manual operator, with the substance on which you are to operate, study better the principles of your art, and you will be able to foresee, predict and calculate every thing. It is your ignorance alone which renders your operations a constant groping in the dark, and a discouraging alternation of success and disappointment."

In fact, the operator who proceeds with a perfect knowledge of the principles of his art, and of the results of its application, will never ascribe the failure of his process to caprice, but will impute it to the neglect of some indispensable precaution in the application of his principle; and his disappointment will serve as a guide for him to calculate better and improve his preparatory process. Convinced that the effects that flow from his principle are invariable, he knows that every kind of loss and damage can proceed only from an error in the application of his principle.

[11] This operation performed on a great scale, that is in a larger boiler, would require too much exactness, as it would be more difficult to command just the due degree of heat in such a boiler than in a small water-bath which may be set on and taken off at pleasure.

[12] That is, of *Reamur*, or 200 of *Fahrenheit*, in like manner, the 80 of Reamur, or boiling point mentioned below, is 212 of Fahrenheit. T.

§ XII.

MILK.

I took twelve *litres* of milk fresh from the cow; I condensed it in the water-bath and reduced it to two-thirds of its volume, frequently skimming it. Then I strained it through a boulting cloth. When cold I took from it the skim which had risen while it was cooling, and bottled it, with the usual process, and afterwards put it in the water-bath which I let boil for two hours; and at the end of several months, I perceived that the cream had separated itself and was swimming in the bottle in the form of flakes. To obviate this inconvenience, I made a second experiment on a like quantity of milk which I condensed in the water-bath, reducing it to one half, instead of one third, as I had done the former. I then added to the milk, so reduced, the yolks of eight new laid eggs well beaten. Having left the whole thus well mingled half an hour on the fire, I completed the experiment as before. This expedient perfectly succeeded.

The yolk of egg had so completely combined all the particles, that at the end of a year, and even of eighteen months, the milk remained as fresh as when I put it in the bottles. The first also was preserved more than two years. The cream which was in flakes disappeared when put on the fire. Both sustained the boiling alike. From both, butter and whey were afterwards obtained. In the different experiments and chymical analyses to which they were exposed, it was found that the last, being much the better, was equal to the best cream sold at Paris to drink with coffee.

§ XIII.

CREAM.

I took five *litres* of cream taken with care from milk of the preceding evening. I condensed it in the water-bath to four *litres*, without skimming it. I took off the skim which was formed above, in order to strain it through a boulting cloth afterwards, and let it cool. After having taken off the skim which had risen while cooling, I put it in half bottles, observing the usual process, and let the water-bath boil for one hour.

At the end of two years this cream was found as fresh as if prepared the same day. I made some good fresh butter with it; making from four to five ounces of butter from half a *litre* of cream.

§ XIV.

WHEY.

I prepared some whey by the ordinary process. When clarified, and grown cold, I put it in bottles, &c. and let it remain in the water-bath which was boiling one hour. However well the whey may be clarified, when put into the water-bath, the application of the heat always detaches some particles of cheese which are deposited. I preserved some in this way two and three years, and before I made use of it, I strained it that it might be very clear. On an emergency you may content yourself with carefully decantering the whey for this purpose.

§ XV.

OF VEGETABLES.

As the difference of climates renders the productions of different countries more or less early, and varies their qualities, kinds and denominations,[13] attention will be given by the operator to the circumstances of the spot in which he resides.

At Paris and its environs, June and July are the best months for preserving green peas (*petits pois verts*), small windsor beans (*petites fèves de marais*), and asparagus (*asperge*). At a later period, these vegetables suffer greatly from heat and dryness. In August and September I preserve artichokes (*artichauts*), French beans (*haricots verts et blancs*), and cauliflowers (*choux-fleurs*).

In general, all vegetables intended to be preserved should be used as recently gathered as possible, and prepared with the utmost rapidity, so that there should be as it were, but one step from the garden-bed to the water-bath.

---

[13] For this reason the translator adds the original names of the vegetables spoken of. It may happen that some of the kinds of fruits and roots mentioned by the author, do not exactly correspond with those which are considered as the same in this country. Whatever peculiarities there may be in the articles themselves, these will hardly affect the treatment they have to undergo in the process of preserving them. T.

§ XVI.

GREEN PEAS.
(PETITS POIS VERTS.)

The *clamart* and the *crochu* are the two kinds of peas which I prefer, especially the latter, which is the most juicy and sweet of all, as well as the earliest, except the *michaux* (hastings), which is the first pea, but this kind is not fit to be preserved. I gather the peas when they are not too young and tender, for they are apt to dissolve in water during the operation. I take them when they are of a middling size. They are then in a more perfect state, and have an infinitely finer taste and flavour. I shell them as soon as they are gathered. I separate the large ones, and they are then put in bottles, the bottles being for that purpose placed on the stool before mentioned, in order that as many peas as possible may, by shaking the bottle, be made to go into them; I then cork the bottles, &c. and put them in the water-bath, which is made to boil for an hour and half, if the season be cool and moist; and two hours in a dry and hot season; and I terminate the operation as before.

I also put in bottles the larger peas which I had separated from those which were more delicate. These, also, I put into the water-bath, which I let boil according to the season, two hours, or two hours and an half.

§ XVII.

ASPARAGUS.
(ASPERGE.)

I clean the asparagus as if for ordinary use, either with the stalk, or the buds only. Before I put them in bottles or jars, I plunge them into boiling water, and afterwards into cold water, in order to take away the peculiar sharpness of this vegetable. The stalks are placed in the jars with great care, the heads being downwards: the buds are put in bottles. After both are well drained, I cork the bottles, &c. and I put them in the water-bath, where they remain only till the water thoroughly boils.

§ XVIII.

WINDSOR BEANS.
(PETITES FÈVES DE MARAIS.)

Neither the *feverole* (the small dried bean) nor the *julienne*, which resembles it, are fit to be preserved. I make use of the genuine Windsor, or broad bean, which is of the thickness and breadth of the thumb, when ripe. I gather it very small, about the size of the end of the little finger, in order to preserve it with its skin. As the skin becomes brown when in contact with the air, I take the precaution of putting the beans in bottles as soon as shelled. When the bottles are full, the beans having been shaken down gently on the stool, and in that way the vacancies in the bottle having been filled up, I add to each bottle a little bunch of savory; I cork them quickly in order to give them one hour's boiling in the water-bath. When this vegetable has been quickly gathered, prepared and preserved, it has a white, greenish colour; on the contrary, when the operation has been tardy, it becomes brown and hard.

§ XIX.

PEELED WINDSOR BEANS.
(FÈVES DE MARAIS DÉROBÉES.)

In order to preserve Windsor beans stripped of their skins, I gather them larger, about half an inch long at the utmost. I take off the skin, bottle them with a small bunch of savory, &c. and I put them in the water-bath, which is made to boil an hour and half.

§ XX.

FRENCH BEANS.
(HARICOTS VERTS ET BLANCS.)

The bean known by the name of *bayolet*, which resembles the Swiss bean, is the kind fittest to be preserved green, with the pod. It combines uniformity with the best taste. I cause the beans to be gathered as for ordinary use. I string them, and put them in bottles, taking care to shake them on the stool, to fill the vacancies in the bottles. I then cork the bottles and put them in the water-bath, which is to boil an hour and half. When the beans are rather large, I cut them lengthways into two or three pieces: and then they do not require being in the water bath longer than one hour.

Of the kinds of haricot, of which the seeds or beans themselves are to be preserved, the *Soissons* haricot is justly entitled to the preference. For want of that, I take the best of any other species of the *haricot blanc* that I can meet with. I gather it when the shell begins to turn yellow. I then shell it immediately, and bottle it, &c. I put it in the water-bath, to give it a two hours' boiling.

§ XXI.

ARTICHOKES.
(ARTICHAUTS.)

To preserve artichokes whole, I gather them of a middling size; after having taken off all the useless leaves and pared them, I plunge them into boiling water, and immediately afterwards into cold water. Having drained them, I put them into jars which are corked, &c. and they receive an hour's boiling.

To preserve cut artichokes (*en quartiers*), I divide them (taking fine specimens) into eight pieces. I take out the choke and leave very few of the leaves. I plunge them into boiling water, and afterwards into fresh water. Having been drained, they are then placed over the fire in a saucepan, with a piece of fresh butter, seasoning, and fine herbs. When half dressed, they are taken from the fire and set by to cool. They then are put in jars, which are corked, tied, luted, &c. and placed in the water-bath, in which they receive half an hour's boiling.

§ XXII.

CAULIFLOWERS.
(CHOUX-FLEURS.)

I plunge the cauliflower, like the artichoke, in boiling water, and then in cold water, after having first plucked it. When well drained, I put it in jars, which are corked, &c. I place it in the water-bath, in order to give it half an hour's boiling, &c.

As the seasons vary, and are sometimes dry and sometimes moist, it will be soon obvious, that it is necessary to study and adapt the various degrees of heat required according to the season. Attention to this circumstance must never be disregarded. For instance, in a cool and damp year, vegetables are more tender and consequently more sensible to the action of fire. In this case, the water-bath should be made to boil seven or eight minutes less; and in dry seasons, when vegetables are firmer, and better support the action of fire, seven or eight minutes boiling should be added.

§ XXIII.

SORREL.
(OSEILLE.)

I gather *oseille* (sorrel), *belle-dame*[14] *noirée* (beet), *laitue* (lettice), *cerfeuil* (chervil), *ciboule* (green onion), &c. in fit proportions. When they have all been well plucked, washed, drained, and minced, I cause the whole to be stewed together in a copper vessel well tinned. These vegetables ought to be well stewed, as if for daily use, and not dried up and burned as is often done in families, when it is intended to preserve them. This quantity of stewing is the most fit. When my herbs are thus prepared, I set them to cool in earthen or stone vessels. Afterwards I put them in bottles with a wide mouth. I cork them, &c. and I put my sorrel in the water-bath, which is allowed a quarter of an hour's boiling merely. This time is sufficient for preserving it ten years untouched, and as fresh as if it was just taken from the garden. This mode is, without dispute, the best and most economical for families and hospitals, civil and military. It is, above all, most advantageous to the Navy: for sorrel thus prepared may be brought from the Indies, as fresh and savoury as if dressed the same day.

---

[14] A species of the *Bella-donna* very generally made use of as an ingredient in French soups. T.

§ XXIV.

SPINAGE, SUCCORY, AND OTHER HERBS.
(EPINARDS ET CHICORÉES.)

Sorrel and succory are prepared as if for daily use. When fresh gathered, plucked, scalded, cooled, squeezed and minced, I put them in bottles, &c. to give them a quarter of an hour's boiling in the water-bath, &c.

Carrots, cabbages, turnips, parsnips, onions, potatoes, celery, chardoons, (*cardons d'Espagne*), red beet, and, generally, all vegetables, may be preserved alike, either simply scalded, or prepared with soup, in order to be used when taken out of the vessel. In the first case, I cause the vegetables to be scalded and half boiled in water with a little salt. I then take them from the water in order to strain them and let them cool; and afterwards put them into bottles, and into the water-bath. I let the carrots, cabbages, turnips, parsnips, and red beet, remain in the water-bath while it boils one hour: and the onions, potatoes, and celery, &c. half an hour. In the other case I prepare my vegetables with soup, either with or without meat, as for ordinary use. When three-fourths boiled and well prepared and seasoned, I take them from the fire to let them cool. Then I put them in bottles, &c. and give them a good quarter of an hour's boiling in the water-bath.

§ XXV.

A SOUP CALLED JULIENNE.

I compose a *Julienne* of carrots, leeks, turnips, sorrel, French beans, celery, green peas, &c. These I prepare in the ordinary way, which consists in cutting the carrots, turnips, leeks, French beans and celery into small pieces, either round or long. Having well plucked and washed them, I put these vegetables into a saucepan over the fire, with a largish piece of fresh butter. When these are half-done, I add the sorrel and green peas. After the whole has been stewed down, I moisten the vegetables with good gravy, prepared for the purpose, with good meat and poultry. I let the whole boil half an hour. Then I withdraw the fire to let it grow cool; and having put the Julienne into bottles, &c. I let it boil half an hour in the water-bath. Julienne prepared in this way, has been kept by me more than two years.

The Julienne *au maigre* is prepared in the same manner, except that, instead of gravy, I moisten my vegetables, when well dressed, with a clear vegetable soup, either of French beans, lentils, or large green peas, which I have preserved; and I give it in like manner half an hour's boiling in the water-bath.

§ XXVI.

VEGETABLE SOUP.
(COULIS DE RACINES.)

I compose and prepare a vegetable soup in the usual way; I make the soup so rich, that a bottle of the size of a *litre* can supply a dish for twelve persons, by adding two *litres* of water to it, before it is made use of. When it has grown cool, I put it in bottles, to give it half an hour's boiling in the water-bath.

§ XXVII.

LOVE-APPLES.
(TOMATES, OU POMMES D'AMOUR.)

I gather love-apples very ripe, when they have acquired their beautiful colour. Having washed and drained them, I cut them into pieces, and dissolve them over the fire in a copper vessel well tinned. When they are well dissolved and reduced one third in compass, I strain them through a sieve sufficiently fine to hold the kernels. When the whole has passed through, I replace the decoction on the fire, and I condense it till there remains only one third of the first quantity. Then I let them become cool in stone pans, and put them in bottles, &c., in order to give them one good boiling only, in the water-bath.

I have not yet tried any experiments with the flower of the love-apple, but there is no doubt that this new method will furnish means of deriving, at a slight expence, a great value from them also.

§ XXVIII.

HERBS AND MEDICINAL PLANTS.
(PLANTES POTAGÈRES ET MÉDICINALES.)

I filled a bottle with mint (*menthe poivrée*) in branches and full of flowers. I stirred it with a stick to make the bottle hold a greater quantity of it. I corked it well, &c. and gave it a slight boiling in the water-bath. It was perfectly preserved.

The same may be done with all plants to be preserved in bunches. The operator will calculate the degree of heat which it will be necessary to give to the several subjects of his experiment.[15]

---

[15] The mode of extracting the juice of plants by means of water has more or less inconvenience. All those juices which have a principle that is very volatile and easy to evaporate, lose infinitely, even in warm water; much more so therefore, when the heat of the water is raised to a higher degree, and when the plants have been left for a long time in digestion.

Aromatic vegetables are infused, when the object is to preserve the aroma, and not impart to the water the extractive principle which the plant contains. Therefore, tea and coffee are made by infusion. But all the theories ancient and modern, and all the new apparatus employed to seize and hold fast the aroma of the coffee are still very deficient.

Ebullition which is often times resorted to in order to extract the aroma of plants by means of distillation, in spite of all the apparatus made use of for keeping the same closed up, most frequently destroys the nature of the productions.

Not only are the principles extracted by the water injured by this first operation, but they scarcely retain any strength after the evaporation which it is usual to make them undergo, in order to form essences of them. The extract therefore, exhibits nothing but the appearance of the soluble and nutritive principles of vegetable and animal substances; since fire, which is necessary to form an essence by means of evaporation, destroys the aroma and almost all the properties of the substance which contains it.

§ XXIX.

THE JUICES OF HERBS.

I have succeeded in preserving very well the juices of such plants as lettuce, chervil, borage (*bourache*), wild succory (*chicorée sauvage*), water-cresses (*cresson de fontaine*), &c. I prepared and purified them by the usual process, I corked them, &c. in order to give them one boiling in the water-bath.

§ XXX.

FRUITS AND THEIR JUICES.

Fruits and their juices require the utmost celerity in the preparatory process, and particularly in the application of heat to the water-bath.

The fruit which is to be preserved either whole or in quarters, ought not to be completely ripe, because it dissolves in the water-bath. In like manner it should not be gathered either at the commencement or the end of the season. The first and the last of the crop have neither the fine flavour, nor the perfume of those which are gathered in the heighth of the season, that is, when the greater part of the crop of each species is ripe at the same period.

§ XXXI.

WHITE AND RED CURRANTS IN BUNCHES. (GROSEILLES ROUGES ET BLANCHES EN GRAPPES.)

I gather the white and red currants apart, and not too ripe. I collect the finest, and in the finest bunches; and I bottle them, taking care to shake them down on the stool, in order to fill up the vacancies in the bottle. Then I cork them, &c. in order to put them in the water-bath which I am careful to watch closely; and as soon as I perceive it boils, I withdraw the fire rapidly, and a quarter of an hour afterwards draw off the water from the bath by means of the cock, &c.

§ XXXII.

WHITE AND RED CURRANTS, STRIPPED. (GROSEILLES ROUGES ET BLANCHES ÉGRENÉES.)

I strip the white and red currants apart. They are immediately put into bottles, and I conclude the operation with the same attention as in preserving the currants in bunches. I preserve a greater quantity of currants stripped, than in bunches; as the stalks always give a harshness to the currant juice.

§ XXXIII.

CHERRIES, RASPBERRIES, MULBERRIES.
(CERISES, FRAMBOISES, MURES ET CASSIS.)

I gather these fruits before they are too ripe, that they may be less squeezed in the operation. I put them in separate bottles, and shake the bottles gently on the stool. I cork them, &c. and I complete them in the same manner, and with the same care as the currants.

§ XXXIV.

JUICE OF RED CURRANTS.

I gather red currants quite ripe, and squeeze them upon fine sieves. I put into a press the skins which remain upon the sieves, in order to extract all the juice which may be in them, and this I mix with the former juice. I perfume the whole with a little raspberry juice, and I strain this decoction through a sieve finer than those used before. I put the juice in bottles, &c. and expose them to the water-bath, with the same attention as the stripped currants, &c.

I proceed in the same manner with the juice of white currants and barberries (*épines-vinettes*), as well as with that of pomegranates, oranges, and lemons.

§ XXXV.

STRAWBERRIES.
(FRAISES.)

I made a number of experiments on the strawberry, and in various ways, without being able to obtain its perfume. I was forced to have recourse to sugar: in consequence, I squeezed some strawberries, and strained them through a sieve, as if I were about to make ice. I added half a pound of powder sugar, with the juice of half a lemon, to a pound of strawberries. I mixed the whole together, and put the decoction in bottles which I corked, &c. I exposed it to a water-bath till it began to boil, &c. This mode succeeded very well, in every respect, except the colour, which was considerably faded; but that may be supplied.

§ XXXVI.

APRICOTS.
(ABRICOTS.)

For the table, the wild and garden apricot (*l'abricot commun, et l'abricot péche*) both taken from trees standing free in the open air, are the best kinds for preservation: I commonly mingle these two kinds together, because the former supports the latter, which has more sugar in it, and which dissolves more from the action of heat. They may nevertheless be prepared apart, provided the precaution be taken, of letting the garden peach remain a few minutes less in the water-bath than the wild peach. That is, as soon as the water-bath begins to boil, the fire is to be taken away from the garden peach, while the fire may be allowed to remain under the wild peach until the water-bath completely boils.

I gather the apricots when they are ripe, but somewhat firm; when, on being squeezed gently between the fingers, the stone is perceived to detach itself from the fruit. As soon as gathered, I cut them in halves, take out the stone, and peel off the skin with a knife as delicately as possible. I put them into bottles, either in halves or quarters, according to the size of the mouth, and shake them on the stool to fill up the vacancies. I add to each bottle from twelve to fifteen almonds; I cork them and put them into the water-bath to receive one boiling only; and I instantly withdraw the fire with the same precaution as made use of in the preparation of the currants, &c.

§ XXXVII.

PEACHES AND NECTARINES.
(PÊCHES, BRUGNONS.)

The *grosse mignonne* and the *calande* are the two kinds of peach which unite the most flavour and perfume. For want of these, I take the best I can meet with.

I gather the nectarine (*brugnon*) more ripe than the peach, because it supports the heat better: and on the other hand, I leave the skin on it in order to preserve it. Moreover, the same process is observed as in preserving the nectarine, the peach, and the apricot; in every instance watching the water-bath closely, as I do in preserving the bunches of currants.

§ XXXVIII.

PRUNES FROM GREEN GAGES, AND PLUMBS.
(PRUNES DE REINE-CLAUDE ET MIRABELLES.)

I have made prunes of whole green-gages, including the stone and the stalk, as well as of other great plums; and even of *perdrigons* and *alberges*, which succeeded very well. But there are these inconveniences in preserving the largest fruits whole, that few of these large plums can be put into even a large jar, since the vacancies cannot be filled up by shaking the fruit, without altogether crushing them; and that when the heat of the water-bath is applied to them, they shrink, and the jars are found half empty.

In consequence, I have abandoned this too expensive mode, and am accustomed to preserve all these large plums, cut in halves, after having taken out the stone. This is the easiest and most economical manner, corks of a sufficiently large size for large objects being very dear, and also rare, when the cork is very fine; the vessels too which have a narrow or middling neck are more easy to be well corked, and the operation is in consequence more certain. As to the *mirabelle* [a small white plum] and all other small plums, I prepare them with the stone in them, after having taken off the stalk; for they are in this way easier to shake close, and they leave but few vacancies in the bottles. In general, I observe, in the preservation of all these prunes, either whole or cut in halves, the same process, care and attention, which I have pointed out under the head of apricots and peaches.

§ XXXIX.

PEARS OF EVERY KIND.

When the pears are peeled, and cut into quarters, and the pips with their husks are taken out, I put them into bottles, &c. in order to place them in the water-bath. I carefully attend to the degree of heat they have to receive, which, if they are of a kind usually eaten raw, should not be more than sufficient to make the water-bath boil. When the preserve consists of pears usually stewed or boiled, then I let them remain boiling in the water-bath, five or six minutes. Pears which have fallen from the tree require a quarter of an hour's boiling, &c.

§ XL.

CHESNUTS, TRUFFLES, AND MUSHROOMS.
(MARRONS, TRUFFES, ET CHAMPIGNONS.)

I pierce *chesnuts* at the point with the point of a knife, as if I meant to roast them. I put them in bottles, and give them one boiling in the water-bath.

Having well washed and brushed the *truffles* in order to take away all the soil, I cut off the upper part gently with a knife. Then I put them into bottles either whole or in pieces, according to the diameter of the neck. The remainder I put in bottles apart. The whole being well corked, &c. I put them in the water-bath to receive an hour's boiling, &c. It is not necessary to recommend that the truffles should be sound, and recently gathered.

I take *Mushrooms* fresh from the bed, well formed and firm. Having peeled and washed them, I put them in a saucepan on the fire, with a piece of butter or some good olive oil, in order to make them eject their liquor. I leave them on the fire till this liquor is reduced one half. I withdraw them in order to let them grow cool in a pan; after which, I bottle them and give them one good boiling in the water-bath.

§ XLI.

THE JUICE OF THE GRAPE OR MUST.

During the vintage of 1808, I took black grapes, carefully gathered from the vine; after having taken away the rotten and green grapes, and stripped the others from the stalks, I squeezed them upon a fine sieve, and afterwards put into a press the husks which remained on the sieve, in order to extract the remainder of the juice; and then put the produce both of the sieve and the press into one cask. Having let it stand in this state twenty-four hours, I put it in bottles, &c. to give it one good boiling in the water bath. When the operation was completed, I withdrew the bottles from the boiler. The action of the fire had precipitated the little colour which the grape-juice had assumed during the preparation, and it was become very white. I then placed it in my laboratory in a rack as if it had been wine.

I repeated all these experiments on the 10th of September 1809, in the presence of the special commission nominated by his Excellency the Minister of the Interior, composed of the most distinguished persons of the profession.

New experiments which I have begun, as well as others which I purpose to make on various objects, will be detailed in a memoir which I shall publish as soon as I shall be able to speak of their result.

OF THE MODE OF MAKING USE OF THE SUBSTANCES WHICH HAVE BEEN PRESERVED.

§ XLII.

MEAT, GAME, POULTRY, FISH.

Meat which has in the preparatory dressing, as well as the boiling it received in the water-bath, received its due quantity of cooking, will, when it is taken to be used, require only to be properly warmed in order to produce both soup and meat (*potage et bouilli*).

For the sake of greater economy, and to lessen the number of bottles and jars wanted, it is better to make in the first instance a good gravy as already pointed out by me. For both the beef and the gravy need only to be warmed, and by adding one half or two-thirds of water to the gravy a good soup is provided.

In this manner, a bottle containing a *litre* of gravy may, by adding two *litres* of boiling water to it at the moment that it is to be used, and adding a little salt, furnish a dozen good messes. Thus it is easy at a very slight expence to keep a little stock of provisions against an emergency and hot weather, when it is so difficult to procure them, more especially in the country.

All meat, poultry, game, and fish, which have received three-fourths of their dressing in the preparatory process, and the remainder in the water-bath, as already pointed out, may, when taken out of the vessels, be heated to the proper degree in order to be instantly served at table. If, for instance, the substance taken from the bottle or jar, had not received either enough previous dressing, or enough heat from the water-bath; it is immediately put on the fire in order to supply what is deficient. Consequently, when the operator has taken due care in making his preparations, having properly seasoned and dressed them, the use to be made of them afterwards, will at all events be easy and convenient; for on the one hand they will need only to be warmed, and on the other hand, they may, if necessary, be eaten cold.

Substances thus prepared and preserved, do not, as might be imagined, require to be consumed as soon as they are opened. Provisions may be used from a vessel eight or ten days after it has been uncorked,[16] care being taken only to replace the cork as soon as the necessary part of the provision has been taken out. Besides, it is easy to regulate the size of the vessels from one to twenty-five *litres* or more, according to the rapidity of the expected consumption.

---

[16] See the report made to the *Société d'Encouragement pour l'Industrie nationale*, by Mr. Bouriat, in the name of the Committee. Two half-bottles, one of milk, the other of whey, after remaining uncorked from twenty to thirty days, had been re-corked with little cork; nevertheless the two substances retained all their properties.

§ XLIII.

JELLIES MADE OF MEAT AND POULTRY.

A well prepared and preserved jelly, carefully taken in pieces out of the jar may be used to garnish cold dishes, or it may be even dissolved in the water-bath, the vessel containing it being first uncorked; afterwards it may be poured in a dish to congeal again before it is made use of.

Under an infinity of circumstances, a cook may be in want of the substances necessary to make a sauce with. But with the essences of meat, poultry, ham, &c. as well as with a provision of jelly well preserved and prepared, they may be furnished in an instant.

The broth or jelly prepared and preserved as pointed out in page 46 is eaten either cold as it is found in the bottles, or diluted with more or less boiling water, in the proportions which persons of experience may judge suitable in the several instances.

§ XLIV.

MILK AND CREAM.

Cream, Milk and Whey, prepared and preserved in the manner already pointed out, are used in the same way, and for the same daily purposes, as the same articles when fresh.

Since cream and milk are perfectly preserved in this manner, there is no doubt that desert-creams might be preserved by a similar process, as well as those which are used for ices. These, having been well prepared and completed before they are put into bottles, will only require to be gently warmed in the water-bath, the bottles being uncorked, in order to facilitate its coming out of the vessel. In this manner creams and ices may be furnished instantly.

§ XLV.
VEGETABLES.

Vegetables put into bottles without being dressed, and entirely submitted to the action of heat in the water-bath, as before described, require to be prepared for use on being taken out of the bottles. This preparation will be made according to the season, and every one's taste and inclination. Attention must be given to the washing of the vegetables when taken from the bottle; and to facilitate the taking them out, I fill the bottle with luke-warm water, and after having drained it of the first water, I wash the vegetables in a second water somewhat hotter, and having drained them, I then prepare them for a meat or vegetable soup.

§ XLVI.

FRENCH BEANS.
(HARICOTS.)

I scald French beans (*haricots verts,*) as if they were fresh in water, with a little salt when not sufficiently dressed by the preserving process. This often happens to them as well as to artichokes, asparagus and cauliflowers. If sufficiently boiled, on being taken out of the bottles, I have only to wash them in hot water in order to prepare them afterwards for vegetable or meat soup.

I scald in the same way the beans of the *haricot blanc*; when sufficiently dressed, I take them from the fire and leave them in the boiling water, half an hour, and even an hour, in order to render them more tender: I then prepare them for soup.

§ XLVII.

PEAS, BEANS, &C.

Green peas are dressed in various ways. If they are ill cooked in the season, it is the cook who is blamed; but if they are not found good in winter, the fault is thrown on the person who has preserved them, though the fault most frequently arises from some of the substances employed; either from the bad butter, or the oil or rancid fat which is made use of through negligence or economy. At another time they are prepared two hours too soon. They are suffered to stick to the bottom of the saucepan when on the fire, and they are served smelling of the butter which is turned into oil with a burnt taste; or they are prepared without care and with too much precipitation. It is thus we see green peas brought to the table swimming in water; but every one has his way. The following is mine.

As soon as the peas have been washed and immediately afterwards drained (for neither this vegetable nor the windsor-bean must be suffered to remain in water, for that would take away their flavour), I put them on the fire in a saucepan with a morsel of good fresh butter. I add to them a bunch of parsley and chives. After having tossed them several times in butter, I dredge them with a little flour, and moisten them immediately afterwards with boiling water up to the level of the peas. I leave them thus to be boiled a good quarter of an hour, until very little sauce remains. Then I season them with salt and a little pepper, and leave them on the fire until they are stewed down; I then take them off the fire immediately, in order to add a piece of fresh butter as large as a nut, with a table spoonful of powder sugar for each bottle of peas. I toss them well without replacing them on the fire, until the butter is melted, and I serve them up in the shape of a pyramid upon a dish, which I take care to warm thoroughly. I have observed several times, that by adding sugar to the peas when upon the fire, and giving them only one boiling, the peas became hard and the sauce ran so that it could no longer bind the peas together. Thus great attention should be given to the not putting in the sugar and the last piece of butter until the moment of serving them up. This is the only way of dressing them well, for neither in summer nor winter ought any sauce to appear among the peas.

There is another mode of eating green peas and which may suit many persons; this consists in simply boiling the peas in water. When done, the water is drained off and the peas are tossed with a piece of good fresh butter, salt, pepper and sugar, all together over a very gentle fire, they are then served up directly upon a very hot dish. Care must be taken that the peas do not boil with the seasoning, otherwise the butter turns into oil, and the green peas are dissolved in the water.

I cook the small windsor-beans, as well with as without their skin, by the same process and with the same attentions which I observe in dressing green peas.

I make an excellent soup-maigre, with large preserved peas which are equally good for a meat soup. As to asparagus, artichokes, cauliflowers, &c. they are dressed in the usual way after having been washed, &c. Green peas, beans, French beans, and all kinds of vegetables may be three-fourths boiled, seasoned at the same time, as is done when intended for immediate use, put into bottles or other vessels when cool, corked, &c. and allowed one half hour's boiling in the water-bath. By these means vegetables will be preserved and quite ready, which may be made use of in an instant, without any other care than to warm them; and there are also many instances in which these vegetables may be eaten cold. In this way all difficulties may be removed in travelling by land or water.

§ XLVIII.

SPINAGE AND SUCCORY.

I dress spinage and succory as usual, in either vegetable or meat soup. Each bottle of a *litre*, contains two or three dishes, either of spinage, or succory according to their strength. When I want to make use of a part only I re-cork the bottle which I keep for another day.

§ XLIX.

VEGETABLE SOUPS.

Having emptied a bottle containing a *litre* of preserved *Julienne*, I add two *litres* of boiling water with a little salt, and I have a dish for twelve or fifteen persons.

As well as a Julienne, a *coulis de racines*, a soup of lentils, carrots, onions, &c. being well prepared, furnishes with the greatest economy, excellent dishes in an instant.

All farinaceous substances, such as oatmeal, rice, spelt, semoulia, vermicelli, and in general every thing that may be formed into a paste, nutritive and easy of digestion, may be prepared and seasoned with either vegetable or meat soup, and even with milk, before they are made to undergo the preserving process, in order to facilitate the use of them at sea and in armies at a moment of necessity.

§ L.

TOMATES AND HERBS.

I use preserved Tomates or *love-apples* in the same manner as those taken fresh in the season. They need only to be properly warmed and seasoned when taken out of the bottle.

A *Sorrel* preserved in the manner pointed out, does not, when taken out of the bottle, in the least differ from fresh sorrel in June. I make use of it in the same way.

As to mint (*menthe poivrée*) and all other plants which may be preserved in bunches; cooks will know how to make the proper use of them, as well as of the juices of herbs.

§ LI.

PRESERVED FRUITS, MARMELADES, &C.

The manner of making use of fruits, preserved by the process I have pointed out, consists, 1st. in putting such fruit into a fruit jar, in the same state in which it is in the bottle, without adding any sugar, because many persons, more especially ladies, prefer fruits with their natural juice. At the same time I prepare another jar with a preserve of grape-syrup or powder sugar, for those who prefer it. I have from experience learnt that grape-syrup preserves the aroma and agreeable acidity of fruits, much better than sugar. This is a very simple and economical mode of preparing an excellent dish of preserved fruits, which is the more convenient, as every one can satisfy his own taste as to the mixture of sugar with his fruits.

2. In order to make preserves with sugar (*compotes sucrées*), I take a pound of preserved fruits, it matters not which; this, on being taken out of the bottle, I put, with its juice, over the fire in a skimming pan, mixing with it four ounces of grape syrup. As soon as it begins to boil, I withdraw it from the fire, and take off the froth by means of a piece of brown paper, which I apply to the surface. As soon as I have skimmed it, I take the fruit gently off the syrup, in order to put it into a fruit-jar. After having reduced the syrup one half over the fire, I put it upon the fruit in the jar. Fruits thus preserved are sufficiently sweet, and have as fine a flavour as a preserve made in the season with fresh fruits.

3d. In order to preserve in brandy either cherries, apricots, green-gages, pears, peaches, mirabelles, &c. (*compotes à l'eau de vie*), I take a pound of preserved fruit, together with its juice, which I put in a saucepan, on the fire, together with a quarter of a pound of grape syrup. When ready to boil, I skim it; after which, I gently take the fruit from off the syrup, and put it in a jar. I leave the syrup on the fire, till it is reduced to one-fourth of its bulk. Then I take it from the fire in order to add to it a glass of good brandy; and having mixed the whole, I pour the hot syrup upon the fruit in the jar, which I take care to close well that the fruit may be better penetrated by the syrup, &c.

The preserved pear and peach may be alike made use of to make a Burgundy wine conserve with cinnamon, as well as *compotes grillées*.

4th. I make a *marmelade*, either of apricots, peaches, green-gages, or mirabelles, by the following process. I take for one pound of preserved fruit, half a pound of grape syrup. I boil the whole together over a quick fire, taking care to stir it well with a spoon to prevent its boiling. When the marmelade is boiled to a slight consistency, I take it off, because the confectionary which is the least boiled is the best. As preserved fruits afford a facility of making confectionary just when it is wanted, they may, by a little boiling only, be had at any time, fresh and of excellent quality.

§ LII.

CURRANT JAM.

The mode of making currant jam with the juice of this preserved fruit, is quite simple. I put half a pound of sugar to one pound of currant juice, which ought to be perfumed with a little raspberry. Having clarified and dissolved my sugar, I put the currant juice to it, and give it three or four boilings; and when it falls from the skimmer in small lumps not larger than a lentil, I take it from the fire to put it in jars, &c.

§ LIII.

SYRUP OF CURRANTS.

In order to make syrup of currants, I warm the juice of this fruit till it is ready to boil. I then strain it through a cloth. By these means I obtain the juice, limpid, and freed from its mucillage. When strained, I add half a pound of grape syrup to a pound of fruit, and put the whole on the fire together; when boiled to the consistence of a slight syrup, I take it from the fire to put it in bottles when it is cold.

There is a very simple and economical mode of making use, not only of currant juice, but that of all fruits which are employed to compose an acid beverage.

This mode consists merely in putting into a glass of water slightly sweetened with grape syrup, a table spoonful of the juice of preserved currant, or of any other fruit that may be at hand, which is poured into another glass and then drank off. This mode is the more convenient, because it will be always easy to have these preserved juices at hand, or to procure them at a small expence. In this way my family has been, for the space of fifteen years, in the habit of making use of currant juice; and most frequently we prepare this substitute for lemonade, without either sugar or syrup.

§ LIV.

ICES.

I have prepared and made, in the mode usually employed in the fruit season, ices of currants, raspberries, apricots, and peaches, as well as strawberries, preserved in the manner pointed out by me.

I made these experiments before the late improvement in the art of making grape syrup, but now that this production has been brought nearly to perfection, the syrup of the acid grape manufactured by Mr. Privat of Meze, will soon advantageously supply the place of the juice of the sugar cane, in the preparation of the ices of fruit. As I have already observed, the grape syrup preserves the aroma of all fruits better than sugar. Sugar overpowers to so great a degree the taste of the fruits, that it is necessary to add some lemons to the ices of fruits, in order, as it were, to bring out the aroma. When therefore the juice of an acid grape shall be made use of, the lemons will become unnecessary, and the ices of fruit will be the richer. The sweet syrups of the grape will be successfully made use of with all ice-creams.

§ LV.

CORDIALS.
(LIQUEURS.)

I have composed liqueurs and ratafies with the juice of preserved fruits and sweetened with grape syrup. These preparations yielded in nothing to the best home-made liqueurs.

The simple and easy modes which I have pointed out, of preparing every kind of preserved fruit for daily use, prove sufficiently that this method, as sure as it is useful, will introduce the greatest economy in the consumption of the produce of the sugar-cane. The consumer, and more especially the manufacturer, who is obliged to lay in during summer, a considerable stock of this foreign commodity for syrups, liqueurs and confectionary, as well as all the objects of pharmacy, may dispense with it; for it will be sufficient if they lay in an adequate stock of fruit in the season, and prepare it in the manner pointed out, to be exempt from the necessity of preparing it with sugar, except on an emergency, and in the quantities actually wanted. It will follow that the greater part of all these fruits will be preserved, altogether without, or at least with a small quantity of sugar; that many of them will be prepared with grape syrup, and that the sugar from the cane will be made use of only for indispensable objects, or to comply with the old habits, and gratify the luxury of a few.

It will follow, that in a plentiful year, sugar will not be necessary in order to provide against a scanty season; and that, with a slight expence, the same enjoyment will be derived from the preserved produce of two, three, and four years, as from a year of plenty.

§ LVI.

CHESNUTS, TRUFFLES, MUSHROOMS.

On taking the chesnuts out of the vessel in which they have been preserved, I plunge them in cold water, sprinkle them with a little fine salt, and roast them in a pan over a quick fire. In this manner they are excellent. The moistening them and the putting salt upon them may be dispensed with, but they must always be roasted over a quick fire.

I make the same use of preserved truffles and mushrooms, as of those recently gathered.

§ LVII.

GRAPE JUICE, OR MUST.

When I made my first experiments of preserving grape juice in its fresh state, I was unacquainted with Mr. Parmentier's "*Information concerning the means of furnishing a substitute for sugar, in the principal uses made of it in medicine and domestic economy.*"[17] It is this valuable information which supplied me with the means of availing myself of fresh experiments, and making use of two hundred bottles of grape-juice preserved by me six months before.

1st. I made very good grape syrup, following the process of Mr. Parmentier, which is literally as follows.

PREPARATION OF GRAPE SYRUP.

"You take twenty-four [French] pints of grape juice and put one half of it in a boiler placed on the fire, with the precaution of not suffering it to boil with too much force. You add fresh juice as that in the boiler evaporates; you skim it and stir the surface, to add to the evaporation. When the whole of the juice has been put into the boiler, you skim it, you take the boiler off the fire, and add some lye-ashes tied up in a cloth, or whiting (*blanc d'Espagne*, Spanish, or Troy-white), or chalk reduced to a powder, and first diluted in a little grape juice, until it no longer effervesces, or, as it were, boils in the liquor which was shaken.

"By these means, the acid contained in the grape, is separated and neutralized. In order to try the liquor, put blue paper into it, and when it does not turn red, then you may be satisfied that the liquor is no longer acid. Replace the boiler on the fire, after having let it settle an instant, and put in two whites of beaten eggs. Strain the liquor through a woollen cloth, fixed on a wooden frame of twelve or fifteen square inches, so that it occupies little room; then boil again, and continue the evaporation.

"In order to know whether the syrup be sufficiently condensed, let some drop from a spoon upon a plate: if the drop falls without spirting or spreading, or if when divided, the halves run into each other again but slowly, then you may infer that it has acquired the proper consistency.

"Pour it into an earthen vessel which is not varnished; and when completely cold, transfer it to vessels of a moderate size, neat, dry, and well corked; and placed it in the cellar. A bottle once opened, should not remain long only half

[17] "*L'instruction sur les moyens de suppléer le sucre dans les principaux usages qu'on en fait pour la médecine et l'économie domestique; par M. Parmentier.*"

filled; and when you make use of it, take care to hold the neck downwards.

"It is hardly possible to determine precisely, the quantity of chalk or ashes necessary to be used. Less is required in the South than in the North, but at all events, more than is necessary will do no harm, since it remains upon the straining cloth with the other insoluble salts and the skim.

"If in order to preserve these syrups for a longer time, you were to carry on the boiling too long, you would find yourself mistaken; for the syrup would not fail to chrystalize at the bottom of the vessel, while the body would become thin: on the other hand, if the syrup were not sufficiently evaporated, it would soon ferment. A housekeeper who has made these syrups twice, will have learnt the degree of boiling which ought to be given to the syrup, better than can be taught her by rule."

SYRUPS AND RATAFIES.

With this same syrup, I have prepared preserves, confectionary, syrups and beverages, as well as liqueurs and ratafies of all the kinds of fruit I have spoken of.

2d. I made syrup of the same grape juice and by the same process, except that I boiled the latter but slightly, that is, one quarter less than the former; as I wished to satisfy myself whether it would be preserved by the application of heat to the water-bath, in the way before pointed out. Having prepared my syrup, I put it when cold, into three half bottles; one full, and the other a quarter empty. I corked and sealed the bottles, and let them remain in the water-bath only till it boiled, &c. I remarked no difference in the full and half full bottles, and all three were completely preserved.

3d. I took six pints of preserved grape juice, to which I added two pints of good old proof brandy, and also two pounds of grape syrup, which I had prepared. This preparation which I mixed well, I made use of to compose four kinds of liqueur, by means of infusions of apricot-kernels, mint, orange flower, badian, which I had prepared before: these liqueurs having been well strained, were found very good, and sufficiently sweet.

4th. I took two bottles of preserved grape juice, which I poured into two other fit bottles. I corked and tied these bottles, and left them standing upright ten days. During this interval, the liquor caused its cork to burst, like the best Champagne wine, and mantled in the same way.

5th. I repeated this last experiment in the same manner. At the end of twelve or fifteen days, observing no appearance of fermentation in the bottles, I uncorked them in order to let in the air, and I then put into them a table spoonful of preserved raspberry juice. Having re-corked and sealed them, I let them remain eight days longer upright. At the end of that time, both the white and the red juice (*le blanc et le rosé*) caused the cork to spring out. They mantled completely, and were very agreeable to the taste, particularly the red, perfumed with raspberry.

After these experiments made of the *Massy* grape [in the department of *Seine and Oise*], it is more than probable that in the fine vineyards of the South, infinitely more precious results will be drawn from the making use of this method. Grape juice will be preserved there, in order, by congelation, to reduce it at will, to the consistence of syrup, after having taken away its acid for the sweet syrup; or if the juice be condensed over the fire, the quantity of boiling made use of for condensing the syrup, will, by the operation of heat in the water-bath in any preparatory process, be rendered immaterial for the preservation of the syrup for several years.[18]

By means of this process, which is easy to be put in practice, and of little expence in the execution, syrups may be obtained clear and white (even when produced from black grapes), and of a pure sweetness, free from a certain flavour of molasses and burnt sugar, from which it has not hitherto been found possible to exempt grape syrup, when boiled in the ordinary mode, sufficient for its preservation.

Thus this precious production, preserved in bottles and vessels of every size, may be transported to a great distance and in all seasons, coming from Bergerac, Mèze, and all the manufactories of the South, to improve the produce of our small vineyards, and make all classes of society share in the enjoyment of this useful resource.

---

[18] The original is more precise, and refers to an instrument made use of in France for ascertaining the density of liquors, an Aræometer, which, at least in its application to grape syrup, is unknown in this country: the words are, *"les degrés de cuisson de 25, 30, ou 33 à l'aréométre, devient indifférent pour conserver ces sirups,"* &c. T.

§ LVIII.

GENERAL OBSERVATIONS.

From this detail of experiments, it is obvious that this new method of preserving animal and vegetable substance, proceeds from the simple principle of applying heat in a due degree to the several substances, after having deprived them as much as possible of all contact with the external air.

It might on the first view of the subject be thought that a substance, either raw or previously acted upon by fire, and afterwards put into bottles, might, if a vacuum were made in those bottles and they were completely corked, be preserved equally well with the application of heat in the water-bath. This would be an error, for all the trials I have made have convinced me that the absolute privation of the contact of external air (the internal air being rendered of no effect by the action of heat), and the application of heat by means of the water-bath, are both indispensable to the complete preservation of alimentary substances.

My object is not like that of the Bourdeaux chymists, to disunite the component parts of the animal substance, and obtain the animal jelly in a separate state, as well as the animal fibre, free from its juice, and so made to resemble tanned leather. Neither is it my endeavour to furnish at a great expence, as in the preparation of portable soup, a tenacious paste or glue, better adapted to derange the stomach than to provide it with a salutary nourishment.

My problem is, to preserve all nutritive substances with all their peculiar and constituent qualities. My experiments prove that I have resolved this problem.[19]

[19] Some persons of enlightened understandings, but who have, perhaps, delivered themselves over to the spirit of system and prejudice, have declared themselves against my method, alleging a pretended impossibility. But is it then difficult on the principles of a sound, natural philosophy, to assign a reason for the preservation of alimentary substance by my process? May we not infer, that the application of *caloric*, or heat, to the water-bath, operates in producing a gentle fusion of the constituent fermenting principles, so as to destroy the predominating agency of fermentation? This agency is an essential condition of fermentation, at least of its taking place with a certain promptitude. Further, there is no fermentation without air; this being also excluded by my method, we have two assignable causes for its success; the theory of which appears to flow naturally from the means practically employed.

Indeed, if we refer to any of the methods made use of and any of the experiments and observations of ancient and modern times, upon the preservation of alimentary substances, with which we are acquainted, we shall find that *fire* is every where the principal agent, either in the natural duration, or in the artificial preservation of vegetable and animal substances.

It is to the solution of this problem that I have devoted my fortune and twenty years of labour and meditation. Happy that I have already been able to render service to my fellow citizens and humanity, I rely on the justice, generosity and intelligence of a wise government, which never fails to encourage useful discoveries. That government will perceive that the inventor of this method of preservation could not obtain from the invention itself an indemnification for his labour and expence. The chief importance of this process lies in its subservience to the wants of civil and military hospitals, and particularly of the Navy. It is in these departments of the public service that my process may be employed in a manner advantageous to the state, and it is from them that I may receive the just reward of my labours. I expect every thing from the beneficent views of the minister, and my expectations will not be disappointed.

Fabroni has proved that heat applied to grape juice or must, destroys the fermentation of this *vegeto-animal*, which is pre-eminently *leaven*. *Thenard* has made like experiments on cherries, gooseberries, and other fruits. The experiments of the late *Vilaris*, and of Mr. *Cazalès*, learned chymists at Bourdeaux, who have dried meat by means of stoves, equally prove that the application of heat destroys the agents of putrefaction.

Drying, boiling, evaporating, as well as the caustic and savoury substances which are employed in the preservation of alimentary productions, all serve to shew that caloric in its various modes of application, produces the same effects.

§ LIX.

PRACTICAL REMARKS.

The bottles and other vessels of every kind fit for the preservation of alimentary substances will occasion but a very slight expence at one time. They may be always used again, if care be taken to rince them as soon as they are empty; good corks, string and wire are not expensive. As soon as the method is known, proper bottles and jars will be met with at the manufacturers, corks of every size and properly prepared for use will be furnished by the cork-cutters, as well as iron-wire fit for use.

It will be always adviseable to procure corks before bottles, and in that case no other bottles need be purchased than such as may have necks suited to the size of the corks, for I have been often unable to procure corks of such a size as I could wish.

The glass-houses of *la Garre, Sèves,* and *des Prémontrés* near *Courcy-le-Château,* are already accustomed to the manufactory of corks and jars necessary for the preserving process. I am most satisfied with the latter, which has served me for the last four years.

Good corking depends only on a little practice. It will suffice to cork a dozen bottles with care and exactness, in order to familiarize a person with the method. Every day, wine and liquors are bottled and transported by land and water to the remotest places. Even glass vessels containing from forty to eighty *litres* in measure have been sent to a great distance full of oil of vitriol and other liquids. It will be the same with animal and vegetable productions, preserved in glass bottles or jars, when sufficient care and attention shall be given them. This is the principal thing required. How many rich liquors and other substances would be better preserved which are either lost or spoiled for want of being well corked!

No one will doubt, after all the experiments I have detailed, that the adoption of this new method, which, as may be seen, unites the greatest economy to a perfection unlooked for till the present time, will secure the following advantages.

1. That of considerably diminishing the consumption of sugar, the produce of the cane, and of giving the greatest extension to the manufactories of grape syrup.

2. That of preserving for use in all countries and all seasons, a number of alimentary and medicinal productions, which being very abundant in some places at certain seasons, are therefore wasted, being considered as of no value; while the same substances, under other circumstances, being much wanted, become of

double and even four-fold value; and sometimes cannot be procured at any price, such as butter and eggs.

3. That of procuring for civil and military hospitals, and even for the armies the most valuable assistance, the details of which would be superfluous here. But the great advantage of this method consists principally in its application to the service of the Navy. It will supply fresh and wholesome provisions for his majesty's vessels on long voyages with a saving of more than fifty per cent. Mariners will in case of illness be furnished with broth, various and cooling beverages, vegetables and fruits; in a word, they will be able to partake of a number of alimentary and medicinal substances, which will alone be sufficient to prevent or cure the diseases contracted at Sea, more especially the worst of them all, the scurvy. These advantages eminently merit the public attention when we reflect that salted provisions, and, above all, their bad qualities, have caused the loss of more lives at Sea than shipwrecks and naval engagements.

4. Medicine will find in this method the means of relieving humanity, by the facility of meeting every where, and in all seasons, animal substances, and all kinds of vegetables, as well as their juices, preserved with all their natural qualities and virtues: by the same means it will obtain resources infinitely precious in the production of distant regions, preserved in their fresh state.

5. From this method will arise a new branch of industry, relative to the productions of France, by their circulation through the interior, and the exportation abroad, of the produce with which nature has blessed the different countries.

6. This method will facilitate the exportation of the wine of many vineyards: wine which can scarcely be kept a year, even when not removed from the spot, may hereafter be preserved many years though sent abroad.

Finally, this invention cannot fail to enlarge the domain of Chymistry, and become the common benefit of all countries, which will derive the most precious fruits from it.

So many advantages, and an infinity of others which the imagination of the reader will easily conceive, produced by one and the same cause, are a source of astonishment.

SOCIETY FOR THE ENCOURAGEMENT OF NATIONAL INDUSTRY.

LETTER FROM THE SECRETARY OF THE SOCIETY FOR THE ENCOURAGEMENT OF NATIONAL INDUSTRY

Paris, 7th April, 1809.

THE SECRETARY OF THE SOCIETY FOR THE ENCOURAGEMENT OF NATIONAL INDUSTRY, TO MR. APPERT, PROPRIETOR AT MASSY.

Sir,

I have the pleasure to transmit to you a copy of the Report made to the Société d'Encouragement, *by Messrs. Guyton-Morveau, Parmentier, and Bouriat, on your preserved vegetable and animal substances. Nothing can be added to the judgment passed by the Committee upon your discovery. They announce, that it has not been in their power to make any experiments, either sufficiently exact, or continued for a sufficient length of time, to enable them to verify to what extent the substances prepared by you may be preserved; but what they have themselves observed, suffices to enable them to form an opinion to which they were previously disposed, by the numerous and decisive testimonies which attest your success.*

The Society are of opinion that they are rendering a service to the country and humanity, when they make known so useful a discovery with the eulogies which it merits. Their desire will be accomplished, should their suffrage determine the public to make use of your productions, and so contribute to confer upon you the just rewards of your labours.

Accept, Sir, the assurance of the perfect respect with which I have the honour to salute you.

MATH. MONTMORENCY,
Secretary, &c.

REPORT OF THE SOCIETY

EXTRACT FROM THE PROCÈS-VERBAL OF THE SITTING OF THE COUNCIL OF ADMINISTRATION, WEDNESDAY, 15TH MARCH, 1809.

Report made by Mr. Bouriat, in the name of a Special Committee, on Vegetable and Animal Substances, preserved by MR. APPERT.

The council referred to a committee, consisting of Messrs. Guyton-Morveau, Parmentier, and myself, the examination of vegetable and animal substances presented by Mr. Appert, and preserved by his process, for more than eight months.

These substances were,

1. *Pot-au-feu* [a standing French dish of boiled meat, fowls, &c.]

2. *Consommé*, gravy.

3. Milk.

4. Whey.

5. Green Peas.

6. Small Windsor Beans.

7. Cherries.

8. Apricots.

9. Currant Juice.

10. Raspberries.

Each of these articles was contained in an earthen vessel hermetically sealed, the cork being fastened with iron wire and pitched. Proceeding methodically in our enquiry:

We found in the *pot-au-feu* a jelly tolerably rich, with a piece of beef and two pieces of fowl in the middle. Warming the whole with care, to a suitable degree, the soup was found good, and the meat which was separated from it, very tender, and of an agreeable flavour.

The *consommé* appeared to us to be excellent; and though prepared fifteen months before, there was scarcely any discernible difference between its then state, and what it would have been, if made fresh the same day.

The *milk* was found to be of a yellowish colour, resembling that of colostrum or beestings, more thick, as well as sweeter and more savoury than the ordinary milk: a superiority it derives from the concentration it has undergone. It may be affirmed that milk of this kind, though prepared nine months before, may supply the place of the greater part of the cream sold at Paris. What however will appear more extraordinary is, that this same milk having been put into a pint bottle which was uncorked a month before, to take out a part of it, and re-corked afterwards with little care, was also preserved, having undergone scarcely any change. At first it appeared to have somewhat thickened, but a slight shaking was sufficient to bring back its ordinary liquidity. I present it here in the same bottle, that you may convince yourselves of a fact, which I should have had a difficulty to believe, if I had heard of it only, without having the evidence before me.

The *whey* which we afterwards examined, presented some singular appearances not less astonishing. It had all the transparency of whey recently prepared. Its colour was deeper, it had a stronger taste, and it was somewhat thicker. It underwent a change also with less rapidity, having been exposed to the air at the end of a fortnight; for a bottle opened six weeks ago, occasionally shaken, and ill corked, did not begin to lose its transparency till the end of a fortnight. Its surface at the end of more than a month was covered with a somewhat thick mouldiness, which when carefully taken off, left the remainder still possessing the flavour of whey.

The *green peas* and the *Windsor beans*, boiled with the attention enjoined by Mr. Appert, furnished two excellent dishes, which the remoteness of the usual season of such vegetables appeared to render still more finely flavoured and agreeable.

Whole *cherries*, and *apricots* cut in quarters preserved a great part of the flavour they had when gathered. It is true Mr. Appert was obliged to gather them before they were quite ripe, lest they should lose too much of their figure in the glass jars in which they were preserved.

The *currant* and *raspberry juice* appeared to us to enjoy almost all their qualities. We found the aroma of the raspberry perfectly preserved, as well as the somewhat aromatic acid of the currant. Their colour only was a little faded.

Such were the results on our examining the substances prepared according to Mr. Appert's process, more than eight months, and some of them a year, and fifteen months before; for instance, the whey. We could only receive his statement as to the time of the previous preparation of these articles, as they had been deposited but two months with the Society; but even this shorter period is sufficient to give us a favourable opinion of the author's process. We are the more justified in relying on Mr. Appert's declarations, as persons highly worthy of credit, have by their own experiments, convinced themselves that similar substances may be preserved for more than a year. Mr. Appert forwarded to the Council mere specimens of the articles I have enumerated; but he prepares a still greater variety of alimentary substances. He did not communicate his process to us.

Observations.

The art of better preserving vegetables and animal substances in the state in which nature produces them, has been to a considerable degree the object both of pharmacy and chymistry. To attain that end various means have been employed.

Desiccation, ardent spirits, acids and oils, saccharine and saline substances, &c. have been made use of; but it must be confessed that these means cause many productions to lose a part of their properties, or otherwise modifies them, so that their aroma and flavour are no longer to be recognized. From this point of view, the process of Mr. Appert appears to us preferable, if without having recourse to desiccation he adds no extraneous substance to that he wishes to preserve. There is every reason to believe that his method is by so much the better, as the substances on which he operates are more capable of sustaining so high a temperature without a sensible change.

Several persons of acknowledged merit, have by desire of the prefects in different Seaports, examined Mr. Appert's preparations. It is only necessary to read the reports made by these well-informed persons, in order to be convinced of the excellence of the author's process.

At Brest, for instance, on the 14th of April 1807, the committee named by the Maritime Prefect express themselves as follows:

"It is demonstrated by every thing just said, that all the alimentary substances, in number eighteen, embarked in the *Stationnaire*, December 12, 1806, and disembarked April 13, 1807, and which were examined by a committee for that especial purpose, under the presidency of a commissary of marine belonging to the hospitals, underwent no change while they were on board, and that they were in the same state at the several periods of the embarkation and disembarkation.

"It may be added that Mr. Appert's process for the preservation of the articles examined, has been followed by all the success he had promised himself; and that with improvement, which he considers as very easy, and finding means to diminish the number of vessels employed, these provisions would offer great advantages on board his majesty's and other vessels."

The Committee nominated at Bourdeaux by the Prefect of the Department, assert, positively:

"The detail which we have just given, on the objects prepared by Mr. Appert, will point out to you that they were in a state of perfect preservation; that the means made use of do not depend on the addition of extraneous substances, and that these means are founded on a process invented or improved by Mr. Appert, which do not destroy the perfume or flavour of the subjects submitted to their influence."

Rear-Admiral Allemand wrote a letter to Mr. Appert, of which I subjoin a copy.

"I communicated your letter, Sir, to the Captains, under my orders, and they tasted the day before yesterday the vegetables I purchased of you fourteen months ago, one bottle of which my *maître-d'hôtel* had by accident left in the store-room. As green peas and beans are just beginning to be gathered, the officers actually believed your preserved vegetables to be fresh, so well had you succeeded; they wish to purchase a large quantity of them, as well as soup, fruit, and meat in bottles. I shall also take a considerable quantity for myself at the end of the season.

"I am so well persuaded, Sir, of the infinite advantage which would attend the providing a quantity of articles for the use of the sick on board, that if his Excellency, the Minister of the Marine and Colonies, should do me the honour to ask for my opinion, I shall not hesitate to confirm this my opinion, as well for the sake of the government and of the sick, as of yourself. I shall take the earliest opportunity to speak with him on the subject. Accept the assurance of my high consideration.

"On board the Imperial Ship *le Majestueux*, at anchor off the *Ile d'Aix*.

(Signed) "ALLEMAND."
"7th March, 1807."

Copy of a Letter of Vice-Admiral Martin, Maritime Prefect, to Mr. Appert, at Brest.

"I have received, Sir, your letter of the 27th of last April. According to your desire, I have addressed to his Excellency, the Minister of the Marine and Colonies, a report of the examination of a variety of provisions prepared according to your process.

"I shall neglect no opportunity of making known a discovery which appears to be as useful to the State as it is interesting to seamen. I have the honour to salute you.

"The vice admiral, maritime-prefect,
(Signed) "MARTIN."
"Rochefort, 22d May, 1807."

*   *   *   *   *

It is apparent from these reports, which appear to be almost the same, though made in towns remote from each other, at different periods and by different persons, that the process of Mr. Appert is as certain as it is useful. It affords the means of enjoying throughout the empire, during the whole year, and with great convenience, the productions which belong alone to a part of it, without fearing that they may have undergone any change by their having been transported to a great distance, or from the remoteness of the season of their growth. Merely under this point of view, the advantage appears to be great: and it has not escaped the notice of the poets and amiable writers, who, to amuse themselves, sing the art of cookery. Mr. Appert has repeatedly received from them the most flattering and highly deserved praises.[20]

The process of this manufacturer is not less valuable in the sparing of sugar in the use of fruit; for without the aid of that article, it preserves the juice till the moment of its consumption, when only a small portion needs to be added to the juice;

---

[20] These *poëtes et littérateurs amiables qui chantent pour s'amuser*, are *M. de Berchoux*, author of the charming poem *la Gastronomie*, and the authors of the *Almanac des Gourmands*, an annual publication written with infinite wit and humour, and which enjoys a higher reputation, and more extensive popularity, than any other work of taste published in France, since the revolution. For an account of both these productions the reader is referred to the LITERARY PANORAMA, Vol. VII. pp. 661, and 719. The Almanac of Gluttons, if it be not the standard of poetic genius in *imperial* France, is at least an indication of the direction which talent is now taking: a direction, which the laws and literary police of the Napoleon government will not fail effectually to maintain. T.

double the quantity would have been necessary to *preserve* the same fruit. It may be further added that the flavour and aroma of substances are better preserved by Mr. Appert's process, than by the decoctions usually made use of in order to preserve them with sugar. This will be considered as a very great advantage, when we reflect how prodigious a quantity of this colonial produce is every year employed to preserve the different kinds of fruit and their juices. The establishment of Mr. Appert has not perhaps been duly appreciated by rich capitalists, who might have given it that desirable extension which it will only gradually receive, if the author is abandoned to his own resources.

The success he has already met with, increases his zeal and makes him carry his views further. He promises to transmit, unchanged, the most agreeable productions of our soil beyond the Line. He purposes to multiply the enjoyments of the Indian, the Mexican, and the African, as well as of the Laplander, and to transport into France from remote regions, an infinity of substances which we should desire to receive in their natural state.

The experiments already made on board several vessels, prove that the sick among a crew will be well satisfied with Mr. Appert's preparations, which furnish them with the means of procuring, when necessary, meat and broth of a good quality, milk, acid fruits, and even anti-scorbutic juices; for Mr. Appert assures us he is able to preserve these also.

With respect to the embarkation of meat necessary for a whole crew on a long voyage, a slight difficulty seems to lie in the requisite multiplicity of bottles. But Mr. Appert will, without doubt, find means to obviate this inconvenience, by the choice of vessels less fragile and of a larger size.

Our opinion of the substances preserved by Mr. Appert, and transmitted to our examination, is, that they were all of a good quality, that they may be made use of without any inconvenience; and that the Society owes great praise to the author for having so far advanced the art of preserving vegetable and animal substances. We are happy to render this homage to the zeal and disinterestedness with which he has laboured to attain his end.

When the relations of commerce shall be rendered more easy, Mr. Appert will require nothing beyond his own talents and perseverance, to establish a branch of commerce as useful to himself as to his country; but at the present moment his fellow citizens cannot better recompense his labours, than by employing the produce of his manufactures.

Note. — Mr. Appert desires to preserve his connection with the Society, in order to inform them of the result of the fresh exertions to which he is about to devote himself, on the invitation of your committee.

*   *   *   *   *

The council concurring in opinion with its committee, adopts the present report and its conclusions, and resolves that it shall be inserted in the minutes of the Society.

(Signed) Guyton-Morveau,

Parmentier,
Bouriat.

(A true copy.)
Math. Montmorency, Secretary.

FINIS.

Lector House believes that a society develops through a two-fold approach of continuous learning and adaptation, which is derived from the study of classic literary works spread across the historic timeline of literature records. Therefore, we aim at reviving, repairing and redeveloping all those inaccessible or damaged but historically as well as culturally important literature across subjects so that the future generations may have an opportunity to study and learn from past works to embark upon a journey of creating a better future.

This book is a result of an effort made by Lector House towards making a contribution to the preservation and repair of original ancient works which might hold historical significance to the approach of continuous learning across subjects.

HAPPY READING & LEARNING!

LECTOR HOUSE LLP
E-MAIL: lectorpublishing@gmail.com

www.ingramcontent.com/pod-product-compliance
Lightning Source LLC
LaVergne TN
LVHW090009200726
843493LV00004B/822